INFORMATION TECHNOLOGY

BRIDGE TO THE WIRED VIRTUALITY

Shiva Sukula

Ph.D., A.D.I.T

Ess Ess Publications

4837/24, Ansari Road, Darya Ganj,
New Delhi-110002
Tel. : 23260807, 41563444
Fax : 011-41563334
E-mail: info@essessreference.com
www.essessreference.com

ESS ESS PUBLICATIONS
4837/24, Ansari Road,
Darya Ganj,
New Delhi-110 002.

Tel.: 23260807, 41563444
Fax: 41563334
E-mail: info@essessreference.com
www.essessreference.com

Rs.975/-

First Published - 2008

ISBN: 81-7000-535-3 (10 Digit)
978-81-7000-535-3 (13 Digit)

Published By Ess Ess Publications and printed at Salasar Imaging Systems

Cover Design by W&P Graphics Pvt. Ltd.

PRINTED IN INDIA

Preface

Information technology goes beyond computers. It includes the networks that connect them, electronic databases and other electromagnetic storage of data, and increasingly, with convergence of media, it includes all telecommunications: not just transmission of data, but also video, voice (and audio generally). Progress in information technology continues at a super-exponential growth rate. This has been true for more than one hundred years and over many different generations of computing technology–mechanical, vacuum tube, transistor, integrated circuit. The progress has come about through improvements of existing technologies as well as the discovery of new ways of doing things. In the next ten to twenty years this should generate staggering results. For example, today's personal computers can perform approximately one billion operations per second. In twenty to twenty-five years, given a continuation of the super-exponential growth rate, a personal computer will be able to perform about 10^{15} operations per second. By various estimates, that is the operational power of the human brain. In other words, in the next twenty years, the growth in computing power should correspond to hundreds of millions of years of human evolution. These computers of the future will not think as humans do, but the progress will be staggering nonetheless.

More than fifty years have passed since computers began to be used in business, yet there remains much we don't know about their influence on libraries in general and information services performance in particular. At a broad level, we can't yet say precisely why computerization had little effect on library productivity for four decades and then, in the mid-1990s, suddenly seemed to become the driving force behind a sharp acceleration in information services. Nor can we say with certainty why the recent information services gains have been so unevenly distributed, appearing in certain libraries that have invested heavily in information technology but not in others that have also spent great sums on computer hardware and software. Bridging the gap between geographical and social borders, global information technologies have become mainstream focuses of researchers, scholars and academics within present and previous years. Now more than ever, the information technology landscape is growing with emerging research and new discoveries to expand to all points of the libraries. Profoundly traversing all facets of compound societies, global information technologies implicate and impact information science, political science, international relations, sociology and more.

During this period of time numerous researchers and academicians have developed a variety of techniques, methodologies, and measurement tools that have allowed them to develop, deliver and at the same time evaluate the effectiveness of several areas of library networking. The explosion of these technologies and methodologies have created an abundance of new, state-of-art literature related to all aspects of this expanding discipline, allowing researchers and practicing educators to learn about the latest discoveries in the field of global information technologies.

Due to rapid technological changes that are continually taking place, it is a constant challenge in this discipline to stay abreast of the far-reaching effects of this worldwide expansion, and to be able to develop and deliver more innovative methodologies and techniques utilizing new technological innovation. In order to provide the most comprehensive, in-depth, and recent advances related to this global phenomenon, as well as to offer a single reference source on all conceptual, methodological, technical and managerial issues, as well as the opportunities, future challenges and emerging trends related to global information technologies, the book is pleased to offer on this rapidly growing discipline, in order to empower students, researchers, academicians, and practitioners with a comprehensive understanding of the most critical areas within this field of study.

Information technologies as a discipline has witnessed fundamental changes during the past two decades, allowing information seekers around the globe to have access to information which two decades ago, was inaccessible. In addition to this transformation, many traditional libraries have taken advantage of the technologies offered by the development of information technologies in order to expand and augment their existing programs and practices. This has allowed librarians to serve their customers, employees more effectively and efficiently in the modern virtual world. With continued technological innovations in information and communication technology and with on-going discovery and research into newer and more innovative techniques and applications, the information technologies discipline will continue to witness an explosion of information within this rapidly growing field.

The diverse and comprehensive coverage of information technologies in book will contribute to a better understanding of all topics, research, and discoveries in this developing, significant field of study. Furthermore, the chapters included in this book will be instrumental in the expansion of the body of knowledge in this enormous field, resulting in a greater understanding of the fundamentals while fueling the research initiatives in emerging fields. Technologists see many of the substantial barriers to further advancement. The chapters in this book describe both the evolving technologies and the breakthroughs that may fuel the information age of the future libraries.

Such a discussion is particularly timely today. We are arriving at a turning point in the history of IT in libraries, with the convergence of three important trends that will shape the future. First, libraries are reevaluating their approaches to IT investment and management as the economy emerges from the post-Internet-bubble downturn. Second, the technology industry is in the midst of a restructuring, as vendors reshape their competitive strategies in response to shifts in the marketplace. Third, library policy makers are assessing the broad impact of computers on libraries' performance and productivity, which will lead to crucial decisions about the development of the IT infrastructure throughout the region. Making the right choices in all these areas requires an open exchange of information and views, and it is in that spirit that I offer this book.

✆ Office {2760554
{2760551
Fax 2762838
Resi. 2600066
Fax 2760577

चौ० चरण सिंह विश्वविद्यालय, मेरठ
CH. CHARAN SINGH UNIVERSITY, MEERUT

प्रो० सन्त प्रसाद ओझा
एफ० इन्स्टी० पी० (लन्दन)
कुलपति

Prof. S.P. Ojha
F. Inst. P. (London)
VICE-CHANCELLOR

Foreword

Groundbreaking research has bolstered the library's information service capabilities in the ever-changing Information Technology (IT) and multimedia environment and given a solid foundation from which to build its knowledge-based activities. Information technology is basic element of most of the functions of exceedingly innovative libraries, many of which are world leaders in information services. The research being carried out in information, computer and communication technologies is vital to libraries and information center and their continued growth in the IT and multimedia based information services. Information, computer and communication technologies have applications in many fields, including telecommunications, software design and the semi-conductor industry.

The integration of information technology in library services is a central matter in ensuring quality in the library and information system. There are two equally-important reasons for integrating information technology in library services. Users must be able to get the information in a fast manner with the use of information technology, since all jobs in the society of the future will be dependent on it, and information technology must be used in library services in order to improve its quality and make it more effective.

This book on current trends in IT application in the library and information sector provides an overview of the efforts made by the author with a view to promoting the integration of information technology in the library and information sector. The chapters presented are discussed thoroughly such as: Introduction, Software. Program And Programming Languages, DOS, Introduction to Library Automation, Library Automation Software, Introduction to Digital Libraries, Digital Libraries: Physical Aspects, Internet: Basics, Internet: Application to Libraries, Library Networks, Metadata, Artificial Intelligence, Data Mining, and Expert System.

I hope that the book will provide a basis for a lively and rewarding understanding of current trends in information technology application to libraries. The knowledge on current trends in information, computers and communications technologies also reinforces the excellence which libraries strive for and achieve in information services and demonstrates the foundation which the libraries and information centers have built for future growth in the IT based information services.

I congratulate Dr. (Ms.) Shiva Shukla for coming out with such an excellent work.

Prof. S.P. Ojha

Contents

1
Introduction

Certainly there have been many technological developments in the last 50 years, during which the computer and the Internet have been evolved. In fact, the impact of computer on teaching and learning has been tremendous. Developments in communication technologies have more influence, with the consequence that online discussion and e-resources dominate library. Audio-visual media have been treated more as an icing-on-the-cake than as something at the very heart of learning and likewise their long-suffering support services. While introducing a computer is a machine that is programmed to manipulate symbols. The principal characteristics of a computer system are:

* It responds to a specific set of instructions in a well-defined manner.
* It executes a prerecorded list of instructions i.e., a program.
* It store and retrieves large amounts of data in a quick manner.

Computer is one of the most revolutionary and powerful tools ever developed because of the tremendous changes it has made. At the least, a computer is an electronic device that executes the instructions in a program.

A computer usually performs following four functions:

a. It accepts data throught the input
b. It processes data by the processing
c. It produces output through Output
d. It stores results in the storage

A computer system can be further explained as:

Hardware is the physical parts of the computer.

Software are the programs (instructions) that tell the computer what to do.

Data are individual facts like first name, price, years etc.

Information is that data which has been presented into a meaningful form, like a complete mailing address

Therefore it is evident that computers perform complex and repetitive procedures in quick, precise and reliable. Modern computers are electronic and digital. The actual machinery involved in making a computer (wires, transistors, and circuits) is called hardware; while the instructions and data which are required to run the system are called software.

COMPONENTS

Computers have the following components:

Central processing unit (CPU) : At the heart of every Personal Computer (PC) is the central processing unit (CPU). The CPU plugs into a motherboard which has a lot of other chips and electronics on it. The CPU and other components work together to schedule, compute and control everything that happens in the PC. This component actually executes instructions organized in programs ("software") which tell the computer what to do.

Memory : This is temporary storage used by the CPU to store results of calculations or files brought in from the hard drive. Memory is very fast and volatile which means it loses the information when power is removed. The memory cells are housed in Integrated Circuits (ICs), or chips as they are often called. Enables a computer to store, at least temporarily, data, programs, and intermediate results.

Mass storage device : These devices that retain information magnetically (Hard Disk Drive and Tapes) or optically (CD and DVD). They are not as fast as memory but can store much more data. They do not lose their information when power is removed. Allows a computer to permanently retain large amounts of data and programs between jobs. Common mass storage devices include disk drives and tape drives.

Input device : These devices allow users to direct the action of the PC. The keyboard and mouse enable user to control the computer by giving it directions (input). Usually a keyboard and mouse, the input device is the conduit through which data and instructions enter a computer.

Output device : These devices display or produce results for user. The video display and printer are the primary ones. A display screen, printer, or other device that lets see what the computer has accomplished.

Software : These are instructions, also known also as programs or code. Software is to the CPU, what knowledge is to human brains. The CPU and software work very closely together. When they work properly together, they can they direct and execute activities in the computer system in a productive way.

THE FIVE GENERATIONS OF COMPUTERS

The history of computer development is often referred to the different generations of computing devices. Each generation of computer is characterized by a major technological development that fundamentally changed the way computers operate, resulting in increasingly smaller, cheaper, more powerful and more efficient and reliable devices.

First Generation - 1940-1956: Vacuum Tubes

The first computers used vacuum tubes for circuitry and magnetic drums for memory and were often enormous, taking up entire rooms. They were very expensive to operate and in addition to using a great deal of electricity, generated a lot of heat, which was often the cause of malfunctions. First generation computers relied on machine language to perform operations, and they could only solve one problem at a time. Input was based on punched cards and paper tape, and output was displayed on printouts. The (UNIVAC) and (ENIAC) computers are examples of first-generation computing devices.

Second Generation - 1956-1963: Transistors

In the second generation, transistors replaced vacuum tubes. The transistor was invented in 1947 but witnessed widespread use in computers in the late 50s. The transistor was far superior to the vacuum tube. It allowed computers to become smaller, faster, cheaper, more energy-efficient and more reliable than their first-generation predecessors. Though the transistor still generated a great deal of heat that subjected the computer to damage, it was a vast improvement over the vacuum tube. Second-generation computers still relied on punched cards for input and printouts for output.

Second-generation computers moved from cryptic binary machine language to symbolic, or assembly languages, which allowed programmers to specify instructions in words. High-level programming languages were also being initiated at this time, such as early versions of COBOL and FORTRAN. These were also the first computers that stored their instructions in their memory, which moved from a magnetic drum to magnetic core technology.

Third Generation - 1964-1971: Integrated Circuits

The development of the integrated circuit was the hallmark of the third generation of computers. Transistors were miniaturized and placed on silicon chips, called semiconductors, which drastically increased the speed and efficiency of computers. Keyboards and monitors interfaced with an operating system provided interaction which allowed the device to run many different applications at one time with a central program that monitored the memory. Computers for the first time became accessible to a mass audience as they were smaller and cheaper.

Fourth Generation - 1971-Present: Microprocessors

The microprocessor brought the fourth generation of computers, as thousands of integrated circuits were built onto a single silicon chip. The Intel 4004 chip, developed in 1971, located all the components of the computer - from the central processing unit and memory to input/output controls - on a single chip. In 1981 IBM introduced its first computer for the home user, and in 1984 Apple introduced the Macintosh. As these small computers became more powerful, they could be linked together to form networks, which eventually led to the development of the Internet. Fourth generation computers saw the development of GUIs, the mouse and handheld devices.

Fifth Generation - Present and Beyond: Artificial Intelligence

Fifth generation computing devices which are based on artificial intelligence, are still in development. There are some applications, such as voice recognition, that are being used today. The use of parallel processing and superconductors is helping to make artificial intelligence a reality. Quantum computation and molecular and nanotechnology will radically change the face of computers. The goal of fifth-generation computing is to develop devices that respond to natural language input and are capable of learning and self-organization.

APPLICATIONS

Introduction

An application is another word for a program running on the computer. Good application depends on how well it performs the tasks it is designed to do and how easy it is for the user to use. Application involves the user interface- the way the user tells the software what to do and how the computer displays information and options to the user.

1. Text Interface : A text interface was available in the beginning. A text-based interface means typing in all the commands. If user mis-types, user has to backspace to error, which erases what user already typed. Add-on programs were written, so that user could edit what was typed. Modern text interfaces have lots of cool shortcuts and features.

Text Interface with Menus : Improvements arrived with the addition of menus and the use of the arrow keys to move around the screen. This was much better than having to type in all the commands.

Graphical Interface : A graphical user interface (**GUI**) uses small pictures and icons to make it easier for the user. It is more user friendly. The example is from' Windows 95/98. The use of drop-down menus, windows, buttons, and icons is found.

Common features of a graphical interface are as following:

* Window
* Menu
* Button
* Icon

2. Word Processing

There are different kinds of applications, with lots of features. Word processing is the application that is used most often and most widely. Examples of word processing programs: Microsoft Word, WordPerfect, Lotus WordPro, and Open Office Writer. Word processing has replaced the typewriter as the main way words are put on paper. Documents can be revised and corrected before they are ever printed. An existing document can be used as a template, or pattern, for a new one.

Steps to produce a document are as following :

* Create
* Edit

* Format
* Print
* Save

3. Database

A database is a collection of data that user wants to manage, rearrange, and add to later. Database is a good program to use to manage lists that are not entirely numbers, such as addresses and phone numbers, inventories, and membership rosters. With a database user could sort the data by name or by any individual item of information recorded. User can create forms to enter or update or just display the data. User can create reports. Both spreadsheets and databases can be used to handle much the same information, but each is optimized to handle a different type most efficiently. The larger the number of records, the more important the differences are. Examples of databases: MS Access, dBase, FoxPro, Paradox, Approach, Oracle, Open Office Base

4. Communications

The communications programs temporarily connect computers to each other to exchange information. They may use telephone lines or dedicated cables for the connection. These programs are not the same as networking programs where computers are actually linked together all the time. Most communications programs now include many different communication functions in one interface.

Purpose : Transmitting data and messages between computers

Major Advantage : Speed

A communication program includes one or more of the following actions:

* sending and receiving files: FTP (File Transfer Protocol)
* exchanging messages in a group: chat programs
* private text messages: instant messaging
* voice messages

* video conferencing
* phone calls over the Internet

5. Web Pages

To create a web page for the Internet, user has to write **HTML** (Hypertext Markup Language) code to define what is to be displayed on the page, create or acquire the images for the page, and then upload the files to a web server. Following programs are required to create web pages :

* HTML editing program
* Graphics editing program
* Uploading program
* Browser
* Web space

COMPONENTS OF COMPUTER SYSTEM

A computer system is composed of various parts. These are as following:

1. Input Devices

Input devices let user enter information into a PC. The keyboard, mouse and touch-screen display are examples of input devices. Input devices get information into the computer by interrupting the CPU from it's main program, in much the same way that user is interrupted in his daily life.

For example, when user press a key on the keyboard the same thing happens. An interrupt handler (software routine) says 'Hey CPU, stop what you're doing, go over to the keyboard port, read the value of the key the operator just pushed and store this value in memory location X.' . The CPU stops, reads the key, stores the key value in location X, and then goes back to doing what it was doing before the interrupt. This is a simple example. Often an interrupt will cause a PC to run a very complex, time consuming routine to service the interrupt. The difference is that user probably get interrupted infrequently compared to a PC, unless user has a really busy job, like a switchboard operator, where it could be a few times every minute. A CPU can be interrupted hundreds, even thousands of times per second depending on the environment it is running in.

Types of Input

* *Data* : Data is the raw facts given to the computer.
* *Programs* : Programs are the sets of instructions that direct the computer.
* *Commands* : Commands are special codes or key words that the user inputs to perform a task, like RUN "ACCOUNTS". These can be selected from a menu of commands like "Open" on the File menu. They may also be chosen by clicking on a command button.
* *User Response* : User response is the answer to the computer's question, such as choosing OK, YES, or NO or by typing in text.

2. Output Devices

Output devices are used to provide usable output as well as status about the operation of the PC. The video display, printer and speaker are examples of output devices.

Video Display : The video display uses data that the CPU puts into a section of memory called video memory. This data is continuously updated and sent to the display at the right rate and in the right order, so that the screen displays images we can read or interpret.

Printer : Most of user has probably used a printer. The PC sends data that user selects such as a file or picture to the printer, usually through either a USB or parallel port on the PC.

Speakers : Most PCs come standard with external speakers that plug into the sound card. These speakers play back any audio source that PC can send to them such as music, voice and status information (beeps), such as those made during start up.

Input/Output (I/O) : There are some ports on the PC that are used to move data out of as well as back into the PC. They are called I/O ports.

Serial Port : The *serial port* is an I/O port. It connects to external devices that need to communicate with the PC. Scanners and digital cameras are examples of devices that can connect to the serial port. It is called the serial port because only one bit of information passes across it during each transfer.

Parallel Port : The *parallel port* is also an I/O port that is almost always used for output. More often than not a printer connected to it. It's called a parallel port because it moves a byte (8 bits) of data in parallel during each data transfer.

Universal Serial Bus : The *Universal Serial Bus* (USB) is a newer higher speed serial I/O port that eventually replace the serial and parallel ports. It has been appearing in PCs for a relatively short time. Many printers and other external devices are already equipped with USB ports. The USB port on a PC can connect directly to any device with a USB port. It can also connect to a USB Controller which in turn can connect to many devices that have USB ports.

3. Central Processing Unit

The **CPU**, or **Central Processing Unit**, is the part of the computer where work gets done. In most computers, there is one processing chip. **Main Memory** stores the commands that the CPU executes and the results.

ALU

It stands for **Arithmetic/Logic Unit.** This is the part that executes the computer's commands. A command must be either a basic **arithmetic** operation: **+ - * /** or one of the **logical comparisons**: **> < = not =**. Everything else has to be broken down into these few operations. Only one operation is done in each Machine Cycle. The ALU can only do one thing at a time but can work very, very fast.

Applications:

These are the various programs that are currently running on the computer. By taking turns with the Machine Cycle, modern computers can have several different programs running at once. This is called **multi-tasking**. Each open application has to have some data stored in Main Memory, even if the application is on rest break and is just sitting there. Some programs (graphics programs are notorious for this) require a lot of the Main Memory space, and may not give it up even if they are shut down!

Control Unit

This is the part of the computer that controls the **Machine Cycle**. It takes numerous cycles to do even a simple addition of two numbers. Some of the important actions are :

Fetch : get an instruction from Main Memory

Decode : translate it into computer commands

Execute : actually process the command

Store : write the result to Main Memory

INPUT/OUTPUT STORAGE

When user enters new data, the keystrokes must be stored until the computer can do something with the new data. When user wants data printed out or displayed, it must be stored somewhere handy first.

Main Memory

The computer stores the data and commands in main memory that are currently being used. When the computer is turned off, all data in Main Memory vanishes. A data storage method of this type is called **volatile** since the data "evaporates." The CPU can fetch one piece of data in one machine cycle

Operating System

This is the instructions that the computer uses to tell itself how it "operates". It's the answer to "Who am I and what can I do?"

Some common examples of operating systems are DOS, various versions of Windows, OS/2, UNIX, LINUX, System 7. These all behave in very different ways and have different hardware requirements.

Unused Storage

It is not that there is always some storage space that is not in use. If space runs out in Main Memory, the computer will **crash**, that is, stop working. There are programs that sense when space is getting short and warn the user. The user should close some of the open applications to free up more space in Main Memory.

Working Storage

The numbers and characters that are the intermediate results of computer operations must be stored until the final values are calculated. These values "in progress" are kept in temporary locations. For example, if the computer is adding up the numbers 1, 2, and 3, it would first add 1 to 2 which yields a value of 3. The 3 is stored in **working storage**. Then the 3 and 3 are added and the new value 6 is stored. The value of 14 is now available to be displayed on the screen or to be printed.

PROCESSING

Machine Cycle

The computer can only do one thing at a time. Each action is be broken down into the most basic steps. The round of steps from getting an instruction back to getting the next instruction is called the Machine Cycle.

The Machine Cycle perorms following actions :

Fetch : get an instruction from Main Memory

Decode : translate it into computer commands

Execute : actually process the command

Store : write the result to Main Memory

Speed

The immense speed of the computer enables it to do millions of steps in a second. The MIPS, standing for millions of instructions per second, is one way to measure computer speeds

Memory Addresses

A method of naming the places is needed where Main Memory stores data. Each location needs a unique name, just like houses in a city need a unique street address. Rather than a street name and house number, memory addresses are just numbers.

A memory address holds 1 byte of data where

1 **bit** = 0 or 1, on or off

1 **byte** = 8 bits

1 **kilobyte** (K or KB) = 1024 bytes

1 **megabyte** (MB) = 1024 kilobytes

It is important to note that computers count by twos and powers of 2. 1024 is 2 × 2 × 2 × 2 × 2 × 2 × 2 × 2 × 2 × 2, that is 2 times itself ten times.

Processor Speed

Many different factors determine how fast his computer gets things done. **Processor speed** is one factor. Processor Speed affected by:

i. System clock rate = rate of an electronic pulse used to synchronize processing (Only one action can take place between pulses.)

Measured in **megahertz (MHz)** where 1 MHz = 1 million cycles per second or **gigahertz (GHz)** where 1 GHz = 1 billion cycles per second.

A computer is a 2.4 GHz machine. It's clock rate is 2.4 billion cycles per.

Bigger number = faster processing

ii. Bus width = the amount of data the CPU can transmit

at a time to main memory and to input and output devices. (Any path bits travel is a bus.)

An 8-bit bus moves 8 bits of data at a time. Bus width can be 8, 16, 32, 64, or 128 so far.

Bigger number = faster transfer of data

Word size = **a word is the amount of data the CPU can** process **at one time.**

An 8-bit processor can manipulate 8 bits at a time. Processors can be 8-, 16-, 32-, or 64-bit so far.

Bigger the number = faster processing

3. Memory

Memory is an integrated circuit (chip) that stores information electronically.

i. Main Memory : In practice, data and instructions are often placed in different sections of memory, but this is a matter of software organization, not a hardware requirement. Also, most computers have special sections of memory that permanently hold programs (firmware stored in ROM), and other sections that are permanently used for special purposes. Main memory (also called *main storage*, or just *memory*) holds the bit patterns of machine instructions and the bit patterns of data. Memory chips and the electronics that controls them are concerned only with saving bit patterns and returning them when requested. No distinction is made between bit patterns that are intended as instructions and bit patterns that are intended as data. The amount of memory on a system is often described in terms of:

kilobyte:	2^{10} = 1024 bytes
megabyte:	2^{20} = 1024 kilobytes
gigabyte:	2^{30} = 1024 megabytes
terabyte:	2^{40} = 1024 gigabytes

TYPES OF MEMORY

a. RAM (Volatile Memory)

This is random access memory. Random Access means that any location in the memory can be read or written immediately. It's like a CD, which is also random access. When user selects a track on a CD it goes to that track and starts playing right away.

Serial Access on the other hand is like an audio or video cassette. User can't access any location user choose to instantly. User might have to rewind or fast forward past lots of stuff to get to the location user want.

i. Dynamic Random Access Memory : DRAM stands for Dynamic Random Access Memory and is the main memory used by the CPU to keep copies of files, programs and other temporary information used by the CPU when it is running. When user sees a computer ad that says '1.2Ghz, 128Mb, 10GB Hard Drive etc,' the 128Mb is the size in mega bytes of the DRAM memory in the PC. A mega byte is 1 million bytes. **Dynamic** memory is a type of memory that must be continuously refreshed in order to retain it's contents. Think of each bit location as if it were a small bucket. The amount of water in the bucket represents the state of the bit stored there. If the bucket is more than half full of water, the bit is a 1. If it is less than half full the bit is a 0. These buckets are leaky (dynamic) due to small holes in the bottom.

DRAM cell leak because the cell structure used is small and inexpensive compared to static (non-leaky) memory. When user charges a cell, setting it to a 1 bit, it immediately begins to lose charge. In a very short time the charge falls below the half way mark and the 1 becomes a 0. Each cell must constantly be refreshed with charge by the CPU or the self- refreshing circuits in newer DRAM devices. The leaky DRAM cells are less expensive than SRAM cells which don't leak.

SIMMs and DIMMs : Up until recently DRAM memory

boards were made in SIMM form, single in-line memory module. Today they are supplied in DIMM form, dual in-line memory module.

ii. Static random access memory : Static random access memory (SRAM) is a type of semiconductor memory. The word "static" indicates that the memory retains its contents as long as power remains applied, unlike dynamic RAM (DRAM) that needs to be periodically refreshed (nevertheless, SRAM should not be confused with read-only memory and flash memory, since it is volatile memory and preserves data only while power is continuously applied). SRAM should not be confused with SDRAM, which stands for *synchronous DRAM* and is entirely different from SRAM, or with pseudostatic RAM (PSRAM), which is DRAM disguised as SRAM. SRAM is a little more expensive, but faster and significantly less power hungry (especially idle) than DRAM. It is therefore used where either speed or low power, or both, are of prime interest. SRAM is also easier to control (interface to) and generally more truly *random access* than modern types of DRAM. Due to a more complex internal structure, SRAM is less dense than DRAM and is therefore not used for high-capacity, low-cost applications such as the main memory in personal computers.

Types of SRAM

1. by transistor type:
   * Bipolar junction transistor used in Transistor-transistor Logic (TTL) and Emitter Coupled Logic (ECL) — very fast but consumes a lot of power.
   * Metal-oxide-semiconductor (MOSFET) used in Complementary Metal-oxide-semiconductor, (CMOS) — low power and very common today.
2. by Function
   * Asynchronous — independent of clock frequency; data in and data out are controlled by address transition.

  - * Synchronous — all timings are initiated by the clock edge(s). Address, data in and other control signals are associated with the clock signals.
3. by feature
   - * ZBT (ZBT stands for zero bus turnaround) — the turnaround is the number of clock cycles it takes to change access to the SRAM from **write** to **read** and vice versa. The turnaround for ZBT SRAMs or the latency between read and write cycle is **zero.**
   - * syncBurst (syncBurst SRAM or synchronous-burst SRAM) — features synchronous burst write access to the SRAM to speed up write operation to the SRAM

Bits and bytes

A bit is a binary digit and is the smallest unit of information in a PC. Bits can only have one of two states, either on or off. These two states represent the two values a bit can have, either 1 or 0. A byte is group of 8 bits. Everything in a computer is stored and calculated using this two value number system, which is called a binary. All storage and calculation operations in a PC are done using binary arithmetic. The reason there are only two states for a bit is because of the nature of transistors, which are the building blocks for all digital electronic circuits. In contrast, the number system we use every day is decimal and it has ten states or values. The decimal digits 0,1,2,3,4,5,6,7,8 and 9 represent the ten values a number can have.

b. Read-only memory (Non-Volatile)

Read-only memory (usually known by its acronym, ROM) is a class of storage media used in computers and other electronic devices. Because data stored in ROM cannot be modified (at least not very quickly or easily), it is mainly used to distribute firmware (software that is very closely tied to specific hardware, and unlikely to require frequent updates).Modern semiconductor ROM chips are not immediately distinguishable from similar chips like RAM modules, except by the part numbers printed on the package. In its strictest sense, ROM refers only to mask ROM (the oldest type of solid state ROM), which is fabricated with the desired data permanently stored in it, and thus can never be modified. However, more modern types such as EPROM and flash EEPROM can be erased and re-programmed multiple times; they are still described as "read-only memory" because the reprogramming process is generally infrequent, comparatively slow, and often does not permit random access writes to individual memory locations, which are possible when *reading* a ROM.

ROM for data storage : Since ROM cannot be modified, it is really only suitable for storing data which is not expected to need modification for the life of the device. To that end, ROM has been used in many computers to store look-up tables for the evaluation of mathematical and logical functions. This was especially effective when CPUs were slow and ROM was cheap compared to RAM. The use of ROM to store such small amounts of data has disappeared almost completely in modern general-purpose computers. However, Flash ROM has taken over a new role as a medium for mass storage or secondary storage of files and programs, comparable to hard disks. This usage enabled by the development of NAND (Not And) flash, which features high endurance, efficient use of chip area (and thus low cost-per-bit), and high read/write throughput. Although NAND flash developed from earlier ROM-like technologies, its current use bears little resemblance to the original function of ROM.

Types of ROMs

a. Semiconductor based : Classic **mask-programmed ROM** chips are integrated circuits that physically encode the data to be stored, and thus it is impossible to change their contents after fabrication. Other types of non-volatile solid-state memory are as following :

i. **Programmable read-only memory** (PROM), or **one-time programmable ROM** (OTP), can be written to or **programmed** via a special device called a **PROM programmer**. Typically, this device uses high voltages to permanently destroy or create internal links (fuses or antifuses) within the chip. Consequently, a PROM can only be programmed once.

ii. **Erasable programmable read-only memory** (EPROM) can be erased by exposure to strong ultraviolet light (typically for 10 minutes or longer), then rewritten with a process that again requires application of higher than usual voltage. Repeated exposure to UV (Ultra Violet) light will eventually wear out an EPROM, but the **endurance** of most EPROM chips exceeds 1000 cycles of erasing and reprogramming. EPROM chip packages can often be identified by the prominent circular "window" which allows UV light to enter. After programming, the window is typically covered with a label to prevent accidental erasure.

iii. **Electrically erasable programmable read-only memory** (EEPROM) is based on a similar semiconductor structure to EPROM, but allows its entire contents (or selected **banks**) to be electrically erased, then rewritten electrically, so that they need not be removed from the computer (or camera, MP3 player, etc.). Writing or **flashing** an EEPROM is much slower (milliseconds per bit) than reading to a ROM or writing to a RAM (nanoseconds in both cases.

iv. **Electrically alterable read-only memory** (EAROM) is a type of EEPROM that can be modified one bit at a time. Writing is a very slow process and again requires higher voltage (usually around 12 V) than is used for read access. EAROMs are intended for applications that require infrequent and only partial rewriting. EAROM may be used as non-volatile storage for critical system setup information; in many applications, EAROM has been supplanted by CMOS RAM supplied by mains power and backed-up with a lithium battery.

b. Flash memory (or simply **flash**) is a modern type of EEPROM invented in 1984. Flash memory can be erased and rewritten faster than ordinary EEPROM, and newer designs feature very high endurance (exceeding 1,000,000 cycles). Modern NAND flash makes efficient use of silicon chip area, resulting in individual ICs with a capacity as high as 16 GB as of 2007; this feature, along with its endurance and physical durability, has allowed NAND flash to replace magnetic in some applications (such as USB flash drives). Flash memory is sometimes called **flash ROM** or **flash EEPROM** when used as a replacement for older ROM types, but not in applications that take advantage of its ability to be modified quickly and frequently

Storage is similar to memory in that both are used to retain information. Storage devices use magnetic or optical means to retain data where memory uses an electronic circuit. They are non-volatile, which means they do not lose the information they contain when they are powered off. They have greater capacity but are much slower than memory devices. Floppy Disks, Hard Disks and Tapes all store information magnetically. CDs and DVDs retain information optically. Storage is used to retain important information like programs (software) and data files.

Disk Format

All magnetic disks are similarly **formatted**, or divided into areas, called

* tracks
* sectors
* cylinders

The formatting process sets up a method of assigning addresses to the different areas. It also sets up an area for keeping the list of addresses. Without formatting there would be no way to know what data went with what. It would be like a library where the pages were not in books, but were scattered around on the shelves and tables and floors. User'd have a hard time getting a book together. A formatting method allows user to efficiently use the space while still being able to find things.

Tracks : A track **is a circular ring on one side of the disk. Each track has a number.**

Sectors : A disk sector is a wedge-shape piece of the disk. Each sector is numbered. On a 5¼" disk there are 40 tracks with 9 sectors each. On a 3½" disk there are 80 tracks with 9 sectors each. So a 3½" disk has twice as many named places on it as a 5¼" disk. A track sector is the area of intersection of a track and a sector.

Clusters : Cluster is a set of track sectors, ranging from 2 to 32 or more, depending on the formatting scheme in use. The most common formatting scheme for PCs sets the number of track sectors in a cluster based on the capacity of the disk. A 1.2 gig hard drive will have clusters twice as large as a 500 MB hard drive. 1 cluster is the minimum space used by any read or write. So there is often a lot of slack space, unused space, in the cluster beyond the data stored there.

Cylinders : A cylinder is a set of matched tracks.

1. On a double-sided floppy, a track from the top surface and the same # track from the bottom surface of the disk make up a cylinder. The concept is not particularly useful for floppies.
2. On a hard disk, a cylinder is made of all the tracks of the same # from all the metal disks that make up the "hard disk". When user put these all together on top of each other, it is something that looks like a tin can with no top or bottom - a cylinder.
3. The computer keeps track of what it has put where on a disk by remembering the addresses of all the sectors used, which would mean remembering some combination of the cylinder, track, and sector. Where the difference

between addressing methods shows up is in the time it takes for the read/write head to get into the right position. The cylinder method writes data down the disks on the same cylinder. This works faster because each metal platter has a read/write head for each side and they all move together. So for one position of the read/write heads, the computer can put some data on all the platters before having to move the heads to a new position.

What happens when a disk is formatted?

* All data is ***erased***.
* ***Surfaces*** are ***checked*** for physical and magnetic defects
* A ***root directory*** is created to list where things are on the disk.

1. Magnetic Disks : Of the various types of Auxiliary Storage, the types used most often involve some type of **magnetic disk**. These come in various sizes and materials. This method uses magnetism to store the data on a magnetic surface.

Advantages : high storage capacity reliable gives direct access to data.

A **drive** spins the disk very quickly underneath a **read/write head**, which does what its name says. It reads data from a disk and writes data to a disk.

Types of Magnetic Disks

a. Diskette / Floppy Disk : The Floppy Disk Drive installs in one of the external drive bays at the front of the PC case and is secured by machine screws. External means user can access the drive from the outside. It transfers data to and from the motherboard by means of a 34 pin ribbon cable. It stores data magnetically on a removable floppy disk. A pickup arm in the drive floats above the disk surface. Floppy disks hold 1.44 Mbytes, which at one time was a large amount of data. Today many programs and files are much larger than this. In spite of being surpassed in size by CD and DVD, floppy drives are still found on many newer PCs.

Sizes

1. 5¼"
2. 3½"

Both sizes are made of mylar with an oxide coating. The oxide provides the magnetic quality for the disk. The "floppy" part is what is inside the diskette covers - a very floppy piece of plastic (i.e. the mylar)

b. Other Removable Magnetic Media : Several other kinds of removable magnetic media are in use, such as the popular Zip disk. All of these have a much higher capacity than floppy disks. Each type of media requires its own drive. The drives and disks are much more expensive than floppy drives and disks, but then, user is getting much larger capacities.

Hard Disks

These consist of 1 or more metal platters which are sealed inside a case. The metal is one which is magnetic. The hard disk is usually installed inside the computer's case, though there are removable and cartridge types, also. The Hard Disk Drive installs in one of the 3-1/2 inch internal drive bays in the PC. It is secured by machine screws. It is powered by a 4 conductor cable coming from the power supply. Data to and from the motherboard is carried on a 40-pin IDE (Integrated Drive Electronics) cable. Data is stored magnetically on multiple rigid disks that are stacked up like pancakes. The sensors float just a few microns above the rotating disk surface and can read and write data at very high rates. Most commercially available hard drives rotate at 5400 or 7200 RPM (revolutions per minute) which translates to 90 or 120 revolutions per second respectively. The data transfer rate from the drive to the motherboard is 33 Mbytes/second in bursts. Newer drives are capable of higher speeds up to 66 Mbytes/sec.

Capacity of Disks

5¼" floppy - 360 KB or 1.2 MB

3½" floppy - 720 KB or 1.44 MB

Hard disks	**capacity**
early ones	= 20 MB
currently	= 120+ GB
	where **1 GB** = 1 **gigabyte**
	= 1024 MB

2. Optical storage

Optical storage is a term from engineering referring to the storage of data on an optically readable medium. Data is recorded by making marks in a pattern that can be read back with the aid of light. A common modern technique used by computers involves a tiny beam of laser light precisely focused on a spinning disc. There are other means of optically storing data and new methods are in development. Optical storage differs from other data storage techniques that make use of other technologies such as magnetism or semiconductors. The term optical drive usually refers to a device in a computer that can read CD-ROMs or other optical discs.

a. Compact Disk Drive

* The CDD installs in one of the external 5-1/4 inch drive bays in the front of the PC case. It is secured by machine screws. Some manufacturers offer special rail-like systems that mount on the CDD. These allow the drive to be removed from the PC without having to remove any screws.
* Data is stored optically on the surface of the disk. A laser attached to an arm that moves back and forth across near the disk surface and sends light toward the disk surface.
* Smooth areas called lands reflect the light back to a photo diode located near the laser. The reflected light is read as a 1. Areas called pits are where the aluminum has been removed. When the laser light hits these, it is scattered and very little is picked up by the photo diode. The absence of light is read as a 0.

* CDD have become the predominant removable storage media for PCs and can store 700 Mbytes of data.
* Data to and from the motherboard is carried on a 40-pin IDE (Integrated Drive Electronics) cable.
* There are two types of Compact Disk drives available for PCs.
    1. CD-ROM (read only memory) is the older type. As the title implies it can only read CDs. It can read any standard CD and most CD-R type disks. It may be able to read some types of CD-RW disks too.
    2. CD-RW (ReWritable)units can read and write CD-R and CD-RW type disks. It can also read standard CD type disks.

b. Digital Video Disk drive

DVD also known as the Digital Versatile Disk drive, installs in one of the external 5 1/4 inch drive bays in the PC. It is secured by machine screws. It is designed to optically access data stored on a DVD. A laser moves back and forth near the disk surface and accesses data at a very fast rate. A 4-pin cable from the power supply plugs into the CDD and provides power to it. Data between the DVD drive and the motherboard is carried on a 40-pin IDE (Integrated Drive Electronics) cable. A standard DVD stores up to 4.7 Gbytes on one side of the disk. Digitally formatted movies can be stored on a standard DVD in MPEG-2 format.

Accessing Data

The process of accessing data has 4 steps.

1. **Seek**
2. **Rotate**
3. **Settle**
4. **Data transfer**

Total time to transfer a kilobyte

for floppies, 175 - 300 ms

for hard drive,	15 - 80 ms	
new hard drives, (Jan. 2000)	.0146 ms	(66.6 MB per sec).

Clearly, getting data from a hard disk is immensely faster than from a floppy

FUTURE MEMORIES

1. Ferroelectric RAM (FeRAM or FRAM)

FRAM is a registered trademark of Ramtron International Corporation. FRAM is used by Ramtron and its licencees to refer to FeRAM technology developed by Ramtron. FeRAM is the accepted generic acronym for ferroelectric random-access memory. It is a type of non-volatile computer memory. It is similar in construction to DRAM, which is currently used in the majority of a computer's main memory, but uses a ferroelectric layer to achieve non-volatility. Although the market for non-volatile memory is currently dominated by Flash ROM, FeRAM offers a number of advantages, notably lower power usage, faster write speed and a much greater maximum number (exceeding 10^{16} for 3.3 V devices) of write-erase cycles. FeRAM is currently one of several "advanced" non-volatile memory (NVRAM) technologies that are attempting to gain acceptance as an alternative to flash by avoiding its key weaknesses – high program and erase voltages, slow programming speed, write-erase endurance that is limited to $\sim 10^5$ cycles. Compared to its primary competitors among the new NVRAM technologies, MRAM and PRAM, FeRAM is more mature with volume production at Fujitsu beginning in 1999. FeRAMs at 1-Mbit densities were available in high volume in 2006 from both Fujitsu and Ramtron. Limited volume production of a 4-Mbit MRAM began at Freescale Semiconductor in July 2006, while PRAM has not yet entered volume production at any manufacturer. However, none of the new NVRAM come close to offering the high storage densities (now multiple gigabits) nor the low cost per storage bit of flash memory. FeRAM is competitive in specialized niche applications where its properties (low write voltage, fast write speed, and much greater write-erase endurance) give it a compelling advantage over flash memory

2. Magnetoresistive Random Access Memory (MRAM)

It is a non-volatile computer memory (NVRAM) technology, which has been under development since the 1990s. Continued increases in density of existing memory technologies, notably Flash RAM and DRAM kept MRAM in a niche role in the market, but its proponents believe that the advantages are so overwhelming that MRAM will eventually become dominant. Unlike conventional RAM chip technologies, in MRAM data is not stored as electric charge or current flows, but by magnetic storage elements. The elements are formed from two ferromagnetic plates, each of which can hold a magnetic field, separated by a thin insulating layer. One of the two plates is a permanent magnet set to a particular polarity, the other's field will change to match that of an external field. A memory device is built from a grid of such "cells.

3. Phase-change memory

It (also known as **PCM, PRAM, PCRAM, Ovonic Unified Memory** and **Chalcogenide RAM C-RAM**) is a type of non-volatile computer memory. PRAM uses the unique behavior of chalcogenide glass, which can be "switched" between two states, crystalline and amorphous, with the application of heat. PRAM is one of a number of new memory technologies that are attempting to compete in the non-volatile role with the almost universal Flash memory, which has a number of practical problems these replacements hope to address. It is the switching time that makes PRAM, and other replacements for flash memory, most interesting. PRAM's temperature sensitivity is perhaps its most notable drawback, one that may require changes in the production process of manufacturers incorporating the technology. PRAM can offer much higher performance in applications where writing quickly is important, both because the memory element can be switched more quickly, and also because single bits may be changed to either 1 or 0 without needing to first erase an entire block of cells. PRAM's high speed, thousands of times quicker than conventional hard drives, make it particularly interesting in nonvolatile memory roles that are currently performance-limited by memory access speed.

4. SONOS

SONOS (Semiconductor-Oxide-Nitride-Oxide-Semiconductor) is a type of high-performance non-volatile memory, or NVRAM. It is similar to the widely used Flash RAM, but offers lower power usage and a somewhat longer lifetime. SONOS is being developed as one of a number of potential Flash replacements, and is currently used in Cypress Semiconductor's PSoC line of products. Generally SONOS is very similar to Flash, but is, in theory at least, much easier to produce. Flash requires the construction of a very high-performance insulating barrier on the base leads of its transistors, often requiring as many as nine different steps, whereas the oxide layering in SONOS can be more easily produced on existing lines. A memory is constructed by fabricating a grid of SONOS transistors along with a small amount of control circuitry. After storing or erasing the cell, the controller can measure the state of the cell by passing a small voltage across the emitter-collector pair; if current flows the cell must be in the "no trapped electrons" state, which is considered to mean "0". If no current is seen the cell is in the "1" state. The needed voltages are normally about 2 V for the erased state, and around 4.5 V for the programmed state. Additionally, the voltage needed to bias the gate during writing is much smaller than in Flash. In order to write Flash, power is first built up in a separate device known as a charge pump, which multiplies the input voltage to between 9 V to 20 V. This process takes some time, meaning that writing to a Flash cell is much slower than reading, often between 100 and 1000 times slower. The pulse of high power also degrades the cells slightly, meaning that Flash devices can only be written to between 10,000 and 100,000 times, depending on the type. SONOS devices require much lower write power, typically 5 to 8 V, and do not degrade in the same way. SONOS does suffer from an unrelated problem, however, where electrons become strongly trapped in the ONO later and cannot be removed again. Over long usage this can eventually lead to enough trapped electrons to permanently set the cell to the "1" state, similar to the problems in Flash. However, in SONOS this requires on the order of a 100,000,000 write cycles, 1000 to 10,000 times better than Flash.

5. Resistive Random Access Memory (RRAM)

It is a new non-volatile memory type being developed by Sharp, Samsung, Fujitsu, Spansion, Macronix, Winbond and other companies. Different forms of RRAM have been disclosed, based on different dielectric materials, spanning from perovskites to transition metal oxides to chalcogenides.

The basic idea is that a dielectric, which is normally insulating, can be made to conduct through a filament or conduction path formed after application of a sufficiently high voltage. The conduction path formation can arise from different mechanisms, including defects, metal migration, etc. Once the filament is formed, it may be reset (broken, resulting in high resistance) or set (re-formed, resulting in lower resistance) by an appropriately applied voltage. Recent data suggest that probably many current paths, rather than a single filament, are involved.

6. Nano-RAM

It is a proprietary computer memory technology from the company Nantero. It is a type of nonvolatile random access memory based on the mechanical position of carbon nanotubes deposited on a chip-like substrate. In theory the small size of the nanotubes allows for very high density memories. Nantero also refers to it as **NRAM** in short. Nantero's technology is based on a well-known effect in carbon nanotubes where crossed nanotubes on a flat surface can either be touching or slightly separated in the vertical direction (normal to the substrate) due to Van der Waal's interactions. In Nantero's technology, each NRAM "cell" consists of a number of nanotubes suspended on insulating "lands" over a metal electrode. At rest the nanotubes lie above the electrode "in the air", about 13 nm above it in the current versions, stretched between the two lands. A small dot of gold is deposited on top of the nanotubes on one of the lands, providing an electrical connection, or terminal. A second electrode lies below the surface, about 100 nm away. NRAM has a density, at least in theory, similar to that of DRAM. DRAM consists of a number of capacitors, which are essentially two small metal plates with a thin insulator between them. NRAM is similar, with the terminals and electrodes being roughly the same size as the plates in a DRAM, the nanotubes between them being so much smaller they add nothing to the overall size. However it seems there is a minimum size at which a DRAM can be built, below which there is simply not enough charge being stored to be able to effectively read it. NRAM appears to be limited only by the current state of the art in lithography. This means that NRAM may be able to become much denser than DRAM, meaning that it will also be less expensive, if it becomes possible to control the locations of carbon nanotubes at the scale the semiconductor industry can control the placement of devices on silicon. Additionally, unlike DRAM, NRAM does not require power to "refresh" it, and will retain its memory even after the power is removed. Additionally the power needed to write to the device is much lower than a DRAM, which has to build up charge on the plates. This means that NRAM will not only compete with DRAM in terms of cost, but will require much less power to run, and as a result also be much faster (write speed is largely determined by the total charge needed). NRAM can theoretically reach speeds similar to SRAM, which is faster than DRAM but much less dense, and thus much more expensive.

TYPES OF COMPUTERS

A. According to Computer sizes and power

Computers can be generally classified by size and power as follows, though there is considerable overlap:

Personal computer : A small, single-user computer based on a microprocessor.

Workstation : A powerful, single-user computer. A workstation is like a personal computer, but it has a more powerful microprocessor and, in general, a higher-quality monitor.

Minicomputer : A multi-user computer capable of supporting up to hundreds of users simultaneously.

Mainframe : A powerful multi-user computer capable of supporting many hundreds or thousands of users simultaneously.

Supercomputer : An extremely fast computer that can perform hundreds of millions of instructions per second.

Further elaborations are as following :

i. Supercomputer

Supercomputer is a broad term for one of the fastest computers currently available. Supercomputers are very expensive and are employed for specialized applications that require immense amounts of mathematical calculations (number crunching). For example, weather forecasting requires a supercomputer. Other uses of supercomputers scientific simulations, (animated) graphics, fluid dynamic calculations, nuclear energy research, electronic design, and analysis of geological data (e.g. in petrochemical prospecting). Supercomputers are one of the many types of modern computing machines which are capable of giving very speedy calculations. It is because of their high speed processing ability that Supercomputer systems are used in such works as animation graphics, weather forecasting, nuclear research work, petroleum research works, crypt analysis, molecular modeling and the likes.

The Mainframe machines are primarily used for a number of purposes but supercomputers are so designed to serve a singular purpose.

ii. Mainframe

Mainframe was a term originally referring to the cabinet containing the central processor unit or "main frame" of a room-filling Stone Age batch machine. After the emergence of smaller "minicomputer" designs in the early 1970s, the traditional big iron machines were described as "mainframe computers" and eventually just as mainframes. Nowadays a Mainframe is a very large and expensive computer capable of supporting hundreds, or even thousands, of users simultaneously. The chief difference between a supercomputer and a mainframe is that a supercomputer channels all its power into executing a few programs as fast as possible, whereas a mainframe uses its power to execute many programs concurrently. In some ways, mainframes are more powerful than supercomputers because they support more simultaneous programs. But supercomputers can execute a single program faster than a mainframe. The distinction between small mainframes and minicomputers is vague, depending really on how the manufacturer wants to market its machines. Mainframes are computers that are the most bulky types and with the advancement of technology, they have almost become extinct.

Used for the purpose of research, engineering works, meteorological calculations, graphics and the likes, mainframes today constitutes Unix, Linux, and IBM's z/OS, OS/390. Mainframe systems were manufactured extensively throughout the 1950s and were marketed by IBM, Control Data, General Electric, NCR, UNIVAC, Honeywell, RC, and Burroughs.

The three important Features of mainframe Computers are :

* Mainframes provide for maximum I/O connectivity as they accommodate huge disc farms.

* Mainframes excel in providing maximum I/O band width.
* Mainframe systems provide for good single thread operations.

iii. Minicomputer

It is a midsize computer. In the past decade, the distinction between large minicomputers and small mainframes has blurred, however, as has the distinction between small minicomputers and workstations. But in general, a minicomputer is a multiprocessing system capable of supporting from up to 200 users simultaneously. The Minicomputers were first built in in 1960s and they immediately became a huge success as 40,000 of the minicomputer systems were immediately sold of making the computers hugely available to the general public. With such a successful market possibility many companies stepped in to venture in the minicomputer market. The most successful among these two hundred companies was Digital Equipment Corporation (DEC) that launched the minicomputer models PDP-11 and VAX 11/780.

The Mini computers are a less bulky version of the mainframe computers. Minicomputers are used by business organization. The difference being that it can support the simultaneous working of up to 200 users and is usually maintained in business organizations for the maintenance of accounts and finances. Besides, there are personal computers or microcomputers. The micro computers include Desktop PC s, Laptops, Handheld computers, Tablet PCs, workstations, palmtops, etc. These are the most frequently used computers.

iv. Workstation

It is a type of computer used for engineering applications (CAD/CAM), desktop publishing, software development, and other types of applications that require a moderate amount of computing power and relatively high quality graphics capabilities. Workstations generally come with a large, high-resolution graphics screen, at large amount of RAM, built-in network support, and a graphical user interface. Most workstations also have a mass storage device such as a disk drive, but a special type of workstation, called a diskless workstation, comes without a disk drive. The most common operating systems for workstations are UNIX and Windows NT. Like personal computers, most workstations are single-user computers. However, workstations are typically linked together to form a local-area network, although they can also be used as stand-alone systems. In networking, workstation refers to any computer connected to a local-area network. It could be a workstation or a personal computer.

v. The Microcomputer

These are the most frequently used computers better known by the name of "Personal computers". This is the type of computer meant for public use. Other than Desktop Computer the choice ranges as follows:

* Personal Digital Computer
* Tablet PC
* Towers
* Work Stations
* Laptops
* Hand Held Computer

PERSONAL COMPUTER

It can be defined as a small, relatively inexpensive computer designed for an individual user. In price, personal computers range anywhere from a few hundred pounds to over five thousand pounds. All are based on the microprocessor technology that enables manufacturers to put an entire CPU on one chip. Businesses use personal computers for word processing, accounting, desktop publishing, and for running spreadsheet and database management applications. At home, the most popular use for personal computers is for playing games and recently for surfing the Internet.

Personal computers first appeared in the late 1970s. One of the first and most popular personal computers was the Apple II, introduced in 1977 by Apple Computer. During the late 1970s and early 1980s, new models and competing operating systems seemed to appear daily. Then, in 1981, IBM entered the fray with its first personal computer, known as the IBM PC. The IBM PC quickly became the personal computer of choice, and most other personal computer manufacturers fell by the wayside. P.C. is short for personal computer or IBM PC. One of the few companies to survive IBM's onslaught was Apple Computer, which remains a major player in the personal computer marketplace. Other companies adjusted to IBM's dominance by building IBM clones, computers that were internally almost the same as the IBM PC, but that cost less. Because IBM clones used the same microprocessors as IBM PCs, they were capable of running the same software. Over the years, IBM has lost much of its influence in directing the evolution of PCs. Therefore after the release of the first PC by IBM the term PC increasingly came to mean IBM or IBM-compatible personal computers, to the exclusion of other types of personal computers, such as Macintoshes. In recent years, the term PC has become more and more difficult to pin down. In general, though, it applies to any personal computer based on an Intel microprocessor, or on an Intel-compatible microprocessor. For nearly every other component, including the operating system, there are several options, all of which fall under the rubric of PC

Today, the world of personal computers is basically divided between Apple Macintoshes and PCs. The principal characteristics of personal computers are that they are single-user systems and are based on microprocessors. However, although personal computers are designed as single-user systems, it is common to link them together to form a network. In terms of power, there is great variety. At the high end, the distinction between personal computers and workstations has faded. High-end models of the Macintosh and PC offer the same computing power and graphics capability as low-end workstations by Sun Microsystems, Hewlett-Packard, and DEC.

MODERN TYPES OF COMPUTER

There are various types of computers. Most of these are according to the size, expected use or capability of the computer. While the term computer can apply to virtually any device that has a microprocessor in it, most people think of a computer as a device that receives input from the user through a mouse or keyboard, processes it in some fashion and displays the result on a screen.

PC : The personal computer (PC) defines a computer designed for general use by a single person. PCs were first known as microcomputers because they were a complete computer but built on a smaller scale than the huge systems in use by most businesses.

Desktop : A PC is not designed for portability. The expectation with desktop systems is that user will set the computer up in a permanent location. Most desktops offer more power, storage and versatility for less cost than their portable brethren.

Laptop : Also called notebooks, laptops are portable computers that integrate the display, keyboard, a pointing device or trackball, processor, memory and hard drive all in a battery-operated package slightly larger than an average hardcover book.

Laptops are completely portable, and use less power and make less noise than desktop models. Laptops are also more expensive than desktops. The price gap is closing, though — laptop prices are falling faster than desktop prices, and laptop PCs actually outsold desktop models.

The Laptops have the same basic hardware, software and operating systems. A desktop computer includes a motherboard, video card, hard drive and other components in a large case. The monitor, keyboard, and other peripherals connect wirelessly or with cables. Whether the case sits vertically or horizontally, it has lots of space for add-in cards, cables and air circulation. A laptop, however, is much smaller and lighter than even the most compact PC tower. Its screen is an integrated part of the unit, as is its keyboard. Instead of a spacious case with lots of room for air circulation, a laptop uses a small, flat design in which all the pieces fit together snugly.

Laptop's inherent portability, components :

* Fit into a compact space
* Conserve power
* Produce less heat than desktop components

Palmtop : More commonly known as Personal Digital Assistants (PDAs), palmtops are tightly integrated computers that often use flash memory instead of a hard drive for storage. These computers usually do not have keyboards but rely on touchscreen technology for user input. Palmtops are typically smaller than a paperback novel, very lightweight with a reasonable battery life. A slightly larger and heavier version of the palmtop is the handheld computer.

Workstation : A desktop computer that has a more powerful processor, additional memory and enhanced capabilities for performing a special group of task, such as 3D Graphics or game development.

Server : A computer that has been optimized to provide services to other computers over a network. Servers usually have powerful processors, lots of memory and large hard drives.

Mainframe : In the early days of computing, mainframes were huge computers that could fill an entire room or even a whole floor! As the size of computers has diminished while the power has increased, the term mainframe has fallen out of use in favor of enterprise server. User'll still hear the term used, particularly in large companies to describe the huge machines processing millions of transactions every day.

Minicomputer : Another term rarely used anymore,

minicomputers fall in between microcomputers (PCs) and mainframes (enterprise servers). Minicomputers are normally referred to as mid-range servers now.

Supercomputer : This type of computer usually costs hundreds of thousands or even millions of dollars. Although some supercomputers are single computer systems, most are comprised of multiple high performance computers working in parallel as a single system. The best known supercomputers are built by Cray Supercomputers.

Wearable : The latest trend in computing is wearable computers. Essentially, common computer applications (e-mail, database, multimedia, calendar/scheduler) are integrated into watches, cell phones, visors and even clothing!

BY WAY OF WORKING

i. Analog Computers

Analog Computers and Digital computers are the two kinds of computers, the combination of which is responsible for the invention of the modern Hybrid computers. The analog systems carry out arithmetic and logical operations by manipulating and processing data which you input such as the weight, temperature, voltage, power density, etc. But unlike the digital computer that changes all inputs to binary digits of "1" and "0", the analog computer does not change inputted data to any such sign language.

The analog computer has become obsolete type of computer these days. It is different from a digital computer as it can perform numerous mathematical operations simultaneously. The digital computers, however could not operate the more complex operations. The Analog System is also unique in terms of operation as it utilizes continuous variables for the purpose of mathematical computation. It utilizes mechanical, hydraulic, or electrical energy or operation.

Mathematical Principles Applied For The Data Processing In Analog Computers:

- * Logarithm
- * Summation
- * Exponentiation
- * Calculus
- * Multiplication and Division
- * Inversion

ii. Digital Computers

Digital Computers are that kind of computing machines which work on the principle of binary mathematics. Digital computing machines when clubbed with Analog computers produce Hybrid computers. The digital systems work on the principle of binary mathematics. In binary mathematics all calculations are represented through "0" and "1".

It is often heard about the word "Byte", "Mega Byte", "Kilo Byte". But what is a Byte? A single Byte is a group of 8 numbers or bits represented through a code series comprising of "0" and "1" .The combination of the digital signal can solve a number of mathematical operations. It is interesting to note that the fundamentals of mathematics applied are addition, subtraction, multiplication and division. But the chief problem is that though it can solve mathematical problems at a tremendous speed, it cannot perform bulk operations simultaneously. For this purpose, digital systems by themselves cannot give very precise results. Hence, for greater efficiency, the Digital systems do not find much application

Data Processing in Digital Computers function with the help of logic circuits. A reliable oscillator helps in the proper synchronization of various logic circuits that are responsible for processing data in the digital computers. In better quality computers, the oscillation rate might even be a billion cycles per second. Operating at such high speed, it can process more than billions of logic and arithmetic problems. A digital computer with the help of binary system cannot only perform mathematical problems but also combine the bytes to produce desired graphics, sounds, images and other such commands.

iii. Hybrid computers

These types of computers are, as the name suggests, a combination of both Analog and Digital computers. The Digital computers which work on the principle of binary digit system of "0" and "1" can give very precise results. But the problem is that they are too slow and incapable of large scale mathematical operation. In the hybrid types of computers the Digital counterparts convert the analog signals to perform Robotics and Process control.

OPEN SYSTEMS INTERCONNECTION

The networks in use today are based in some manner on the **Open Systems Interconnection** (OSI) standard. **OSI** was developed in 1984 by the International Organization for Standardization (ISO), a global federation of national standards organizations representing approximately 130 countries. The core of this standard is the **OSI Reference Model**, a set of seven layers that define the different stages that data must go through to travel from one device to another over a network. The **Open Systems Interconnection Basic Reference Model** (*OSI Reference Model* or *OSI Model* for short) is a layered, abstract description for communications and computer network protocol design, developed as part of the Open Systems Interconnection (OSI) initiative. It is also called the **OSI seven layer model**. The layers, are, from top to bottom, Application, Presentation, Session, Transport, Network, Data Link and Physical. A layer is a collection of related functions that provides services to the layer above it and receives service from the layer below it. For example, a layer that provides error-free communications across a network provides the path needed by applications above it, while it calls the next lower layer to send and receive packets that make up the contents of the path.

The newer IETF and IEEE protocols, and OSI protocol work subsequent to the publication of the original architectural standards that have largely superseded it, the OSI model is an excellent place to begin the study of network architecture. This is not always easy to do as many of the protocols in use on the Internet today were designed as part of the TCP/IP model, and may not fit cleanly into the OSI model

In particular, Internet protocols are deliberately not as rigorously architected as the OSI model, but a common version of the TCP/IP model splits it into four layers. The Internet Application Layer includes the OSI Application Layer, Presentation Layer, and most of the Session Layer. The End-to-End Layer includes the graceful close function of the OSI Session Layer as well as the Transport Layer. The Internetwork Layer is equivalent to the OSI Network Layer, while its Interface layer includes the OSI Data Link and Physical Layers.

OSI Model

	Data unit	*Layer*	*Function*
Host layers	Data	7. Application	Network process to application
		6. Presentation	Data representation and encryption
		5. Session	Interhost communication
	Segments	4. Transport	End-to-end connections and reliability (TCP)
Media layers	Packets	3. Network	Path determination and logical addressing (IP)
	Frames	2. Data link	Physical addressing (MAC & LLC)
	Bits	1. Physical	Media, signal and binary transmission

THE LAYERS

The layers are as following:

Application Set

Layer 7: Application – This is the layer that actually interacts with the operating system or application whenever the user chooses to transfer files, read messages or perform other network-related activities. This layer supports application and end-user processes. Communication partners are identified, quality of service is identified, user authentication and privacy are considered, and any constraints on data syntax are identified. Everything at this layer is application-specific. This layer provides application services for file transfers, e-mail, and other network software services. Telnet and FTP are applications that exist entirely in the application level. Tiered application architectures are part of this layer.

Layer 6 : Presentation – Layer 6 takes the data provided by the Application layer and converts it into a standard format that the other layers can understand. This layer provides independence from differences in data representation (e.g., encryption) by translating from application to network format, and vice versa. The presentation layer works to transform data into the form that the application layer can accept. This layer formats and encrypts data to be sent across a network, providing freedom from compatibility problems. It is sometimes called the *syntax layer*.

Layer 5: Session – Layer 5 establishes, maintains and ends communication with the receiving device. This layer establishes, manages and terminates connections between applications. The session layer sets up, coordinates, and terminates conversations, exchanges, and dialogues between the applications at each end. It deals with session and connection coordination.

Transport Set

Layer 4: Transport – This layer maintains **flow control** of data and provides for error checking and recovery of data between the devices. Flow control means that the Transport layer looks to see if data is coming from more than one application and integrates each application's data into a single stream for the physical network. This layer provides transparent transfer of data between end systems, or hosts, and is responsible for end-to-end error recovery and flow control. It ensures complete data transfer.

Layer 3: Network – The way that the data will be sent to the recipient device is determined in this layer. Logical protocols, routing and addressing are handled here. This layer provides switching and routing technologies, creating logical paths, known as virtual circuits, for transmitting data from node to node. Routing and forwarding are functions of this layer, as well as addressing, internetworking, error handling, congestion control and packet sequencing.

Layer 2: Data – In this layer, the appropriate physical protocol is assigned to the data. Also, the type of network and the packet sequencing is defined. At this layer, data packets are encoded and decoded into bits. It furnishes transmission protocol knowledge and management and handles errors in the physical layer, flow control and frame synchronization. The data link layer is divided into two sublayers: The Media Access Control (MAC) layer and the Logical Link Control (LLC) layer. The MAC sublayer controls how a computer on the network gains access to the data and permission to transmit it. The LLC layer controls frame synchronization, flow control and error checking.

Layer 1: Physical – This is the level of the actual hardware. It defines the physical characteristics of the network such as connections, voltage levels and timing. This layer conveys the bit stream - electrical impulse, light or radio signal — through the network at the electrical and mechanical level. It provides the hardware means of sending and receiving data on a carrier, including defining cables, cards and physical aspects. Fast Ethernet, RS232, and ATM are protocols with physical layer components.

PROTOCOL STACKS

A protocol stack is a group of protocols that all work together to allow software or hardware to perform a function. The **TCP/IP** protocol stack is a good example. It uses four layers that map to the OSI model as follows:

Layer 1: Network Interface – This layer combines the Physical and Data layers and routes the data between devices on the same network. It also manages the exchange of data between the network and other devices.

Layer 2: Internet – This layer corresponds to the Network layer. The **Internet Protocol** (IP) uses the IP address, consisting of a **Network Identifier** and a **Host Identifier**, to determine the address of the device it is communicating with.

Layer 3: Transport – Corresponding to the OSI Transport layer, this is the part of the protocol stack where the **Transport Control Protocol** (TCP) can be found. TCP works by asking another device on the network if it is willing to accept information from the local device.

Layer 4: Application – Layer 4 combines the Session, Presentation and Application layers of the OSI model. Protocols for specific functions such as e-mail (**Simple Mail Transfer Protocol, SMTP**) and file transfer (**File Transfer Protocol, FTP**) reside at this level.

The OSI protocols are a family of information exchange standards developed jointly by the ISO and the ITU-T from 1977 onwards. The OSI model describes seven 'layers' of interconnection: the physical layer, the data link layer, the network layer, the transport layer, the session layer, the presentation layer, and the application layer. These OSI protocols for those layers are described below:

1. Physical Layer

This layer deals with the physical plugs and sockets and electrical specification of signals. This is the medium over which the digital signals are to be transmitted. This can be twisted pair, coaxial cable, optical fiber, wireless, or other media.

2. Data Link Layer

The Data Link layer packages raw bits from the Physical layer into frames (logical, structured packets for data). This layer

Examples:

Layer #	*Misc. examples* **Name**	*TCP/IP suite*	*SS7*	*Apple Talk suite*	*OSI suite*	*IPX suite*	*SNA*	*UMTS*
7	Application	NNTP, HL7, Modbus, SIP, SSI	DHCP, DNS, FTP, Gopher, HTTP, NFS, NTP, RTP, SMPP, SMTP, SNMP, Telnet	ISUP, INAP, MAP, TUP, TCAP	AFP	FTAM, X.400, X.500, DAP		APPC
6	Presentation	TDI, ASCII, EBCDIC, MIDI, MPEG	MIME, XDR, SSL, TLS (Not a separate layer)		AFP	ISO 8823, X.226		
5	Session	Named Pipes, NetBIOS, SAP, SDP	Sockets. Session establishment in TCP. SIP. (Not a separate layer with standardized API.)	ASP, ADSP, ZIP, PAP	ISO 8327, X.225	NWLink	DLC?	
4	Transport	NBF, nano TCP, nano UDP	TCP, UDP, SCTP	SCCP	ATP, NBP,	TP0, TP1, AEP, RTMP TP4	SPX TP2, TP3,	

3	Network	NBF, Q.931	IP, ICMP, IPsec, ARP, RIP, OSPF	MTP-3	DDP	X.25 (PLP), CLNP	IPX	RRC (Radio Resource Control) PDCP (Packet Data Convergence Protocol) & Broadcast /Multicast Control (BMC)
2	Data Link	802.3 (Ethernet), 802.11a/b/g/n MAC/LLC, 802.1Q (VLAN), ATM, CDP, FDDI, Fibre Channel, Frame Relay, HDLC, ISL, PPP, Q.921, Token Ring	PPP, SLIP, PPTP, L2TP	MTP-2	LocalTalk, TokenTalk, EtherTalk, AppleTalk Remote Access, PPP	X.25 (LAPB), Token Bus	IEEE 802.3 framing, Ethernet II framing	LLC (Logical Link Control), MAC (Media Access Control)

1	Physical	RS-232, V.35, V.34, I.430, I.431, T1, E1, 10BASE-T, 100 BASE-TX, POTS, SONET, DSL, 802.11a/ b/g/n PHY	MTP-1	RS-232, RS-422, STP, PhoneNet	X.25 (X.21bis, EIA/TIA-232, EIA/TIA-449, EIA-530, G.703)

is responsible for transferring frames from one host to another. It might perform error checking.

3. Network Layer

This layer deals with Connectionless Network Service (CLNS), Connectionless Network Protocol (CLNP) – ISO 8473, Connection-Oriented Network Service (CONS), Connection-Oriented Network Protocol (CONP) – ISO 8878. This is a minor variant of the previously existing X.25 protocol and is compatible with X.25 networks and Network Fast Byte Protocol – ISO 14700

4. Transport Layer

This is specified by ISO 8072 (service) and ISO 8073 (protocol). This layer deals with Transport Protocol Class 0 (TP0), Transport Protocol Class 1 (TP1), Transport Protocol Class 2 (TP2), Transport Protocol Class 3 (TP3), Transport Protocol Class 4 (TP4), and Transport Fast Byte Protocol – ISO 14699.

5. Session Layer

This layer deals with Session service – ISO 8306 / X.215, Connection-oriented Session protocol – ISO 8307 / X.225 and Connectionless Session protocol – ISO 9548.

6. Presentation Layer

This layer deals with Presentation service – ISO 8822 / X.216, Connection-oriented Presentation protocol – ISO 8823 / X.226 and Connectionless Presentation protocol – ISO 9576

7. Application Layer

This layer deals with Common-Application Service Elements (CASEs), Association Control Service Element (ACSE), Remote Operations Service Element (ROSE), Reliable Transfer Service Element (RTSE), Commitment, Concurrency, and Recovery service element (CCRSE) and Security Exchange Service Element (SESE).

8. Application Processes

The application processes which are dealt as following:

* Virtual terminal protocol (VTP)
* Remote Database Access (RDA)
* Distributed Transaction Processing (OSI TP)
* Document Printing Application (DPA)
* Document Filing and Retrieval (DFR)
* Interlibrary Loan Application Protocol (ILAP)
* Document Transfer And Manipulation (DTAM)
* Common management information protocol (CMIP) – ISO 9596 / X.700
* Directory services (DS) – X.500
* File transfer, access, and management (FTAM)
* Message handling system (MHS) – X.400)

9. Routing Protocols

The routing protocols are as following:

* Interdomain Routing Protocol (IDRP) – ISO 10747
* Intermediate System to Intermediate System (IS-IS) – ISO 10589
* End System to Intermediate System (ES-IS) – ISO 9542

REFERENCES

1. www.grassrootsdesign.com/intro
2. www.fayette.k12.il.us/99/Intro2Comp introduction_ to_computers.html
3. www.jegsworks.com/Lessons/lesson1-2/lesson1-1.htm
4. csrc.nist.gov/publications/nistpubs/800-12/handbook.pdf
5. cs244a.stanford.edu/
6. www.sunysb.edu/philosophy/faculty/pgrim/intro.html
7. http://simerr.une.edu.au/resources/researchers/randr1.html
8. en.wikipedia.org/wiki/Computer_hardware
9. www.computerhelpatoz.com/hardware.html
10. www.techtutorials.info/hardware.html

11. www.karbosguide.com
12. www.pccomputernotes.com
13. dir.indiamart.com/industry/computer-hardware.html
14. chortle.ccsu.edu/CS151/Notes/chap01/..%5Cchap01%5Cch01_3.html
15. www.tekxam.com/StudyGuide/concepts/General-Concepts/Computer_Components.html
16. agg3333.ifas.ufl.edu/components.htm
17. www.comptechdoc.org/hardware/pc/begin/hwcomputer.html
18. aspire.cs.uah.edu/textbook/introcomp6.html
19. www.techfest.com/hardware/index.htm
20. www.top-computer-hardware.com
21. en.wikipedia.org/wiki/Computer_types
22. www.cs.cmu.edu/People/fgandon/lecture/uk1999/computers_types
23. computer.howstuffworks.com/question543.htm
24. www.unm.edu/~tbeach/terms/types.html
25. www.elearningplace.co.uk/6852/taster/tastertypescomp.php
26. www.cs.bris.ac.uk/Teaching/Resources/COMS11200/pages/tour7/sld001.htm
27. www.jegsworks.com/Lessons/lesson1-2/lesson1-1.htm
28. www.webopedia.com/TERM/C/memory.html
29. gleez.com/articles/it-systems/computer-memory-ram
30. computer.howstuffworks.com/computer-memory.htm
31. www.memorystore.com
32. www.osdata.com/system/physical/memory.htm
33. www.bitpipe.com/tlist/Memory.html
34. en.wikipedia.org/wiki/OSI_model
35. en.wikipedia.org/wiki/Analog_computer
36. www.play-hookey.com/analog
37. pespmc1.vub.ac.be/ASC/DIGITA_COMPU.html
38. www.britannica.com/eb/article-9030431/digital-computer
39. www.britannica.com/eb/article-9007340/analog-computer
40. pespmc1.vub.ac.be/ASC/ANALOG_COMPU.html
41. www.science.uva.nl/museum/AnalogComputers.html
42. www.webopedia.com/quick_ref/OSI_Layers.asp

43. searchnetworking.techtarget.com/sDefinition/0,,sid7_gci212725,00.html

44. http://download.micron.com/pdf/technotes/ddr2/TN4716.pdf

45. http://ieeexplore.ieee.org/Xplore/login.jsp?url=/iel2/662/5918/00229238.pdf?tp=&arnumber=229238&isnumber=5918

2
Software

The words software, code and programs are often used interchangeably. These refer to instruction sequences that run on the hardware of a computer. Computer software is held in the storage of computers for a very specific purpose and performs the function of the program it implements, either by directly providing instructions to the computer hardware or by serving as an input to another piece of software. Computer software is described as the information processed by the systems. Hardware is any physical device that makes up/plugs into computer. Software is everything else, the programmes that allow user to perform various tasks. It includes therefore word processors, databases, spreadsheets, system utilities, antivirus programmes, email programmes, web browsers and indeed almost everything that allows computer to act as anything other than a paperweight.

The operating system is also software, and usually deals with all questions of hardware, and provides some common routines for the programmes to use. These days, it usually provides a graphical user interface (GUI - note that for the Linux operating system, and for many other free Unices, the GUI is a separate programme that runs on top of the operating system, although it has many operating-system-like qualities). It will also provide many other small programmes, usually diagnostic utilities, and even games.

Basically computer software can be classified into two types:

1. Systems software : It includes basic input-output systems, device drivers etc. In most cases this software comes along in the box when the computer is purchased.

System software is a catch-all term for the programs that handle the running of the computer's hardware. The main categories are:

a. operating systems
b. utility programs
c. network operating systems and
d. database managers

2. Applications software : It includes any data entry, update, query or report program that processes data for the user. It includes the generic productivity software (spreadsheets, word processors, database programs, etc.) as well as custom and packaged programs for payroll, billing, inventory and other accounting purposes.

This type of software includes office suites and video games, and may be purchased separately from the computer. Computer users can also create their own computer software, including spreadsheet templates, word processor macros etc.

OPERATING SYSTEMS

Any desktop or laptop PC normally comes pre-loaded with *Windows XP*. Macintosh computers come pre-loaded with *OS X*. The operating system (OS) is the first thing loaded onto the computer — without the operating system, a computer is useless. Recently, operating systems have started to pop up in smaller computers as well. Operating systems can now be found on many of the devices from cell phones to wireless access points. The computers used in these little devices have gotten so powerful that they can now actually run an operating system and applications. The computer in a typical modern cell phone is now more powerful than a desktop computer from 20 years ago. In any device that has an operating system, there's usually a way to make changes to how the device works.

For a desktop computer user, one can add a new security update, new application or often even a new operating system entirely rather than junk computer and start again with a new one when it is needed to make a change. As long as it is understood how an operating system works and know how to get at it, one can in many cases change some of the ways it behaves. The purpose of an operating system is to organize and control hardware and software so that the device it lives in behaves in a flexible but predictable way. The operating system in desktop computer works and takes control of the other operating systems. The operating system (OS) is essential code that a PC must have. It is stored on hard drive. It provides the basic operating platform that brings the PC to a point where it can interpret the keyboard and mouse input and display data and images on the video display. Other programs can't run on a PC until the OS is loaded and running. DOS, Windows 95 and Windows 98 are examples of operating systems. When an OS like Windows 98 has been loaded and is ready to go, the Windows Desktop appears.

FUNCTIONS OF THE OPERATING SYSTEM

At the simplest level, an operating system does two things

1. It manages the hardware and software resources of the system. In a desktop computer, these resources include such things as the processor, memory, disk space, etc.
2. It provides a stable, consistent way for applications to deal with the hardware without having to know all the details of the hardware.

The first task, managing the hardware and software resources, is very important, as various programs and input methods compete for the attention of the central processing unit (CPU) and demand memory, storage and input/output (I/O) bandwidth for their own purposes. In this capacity, the operating system plays the role of the good parent, making sure that each application gets the necessary resources while playing nicely with all the other applications, as well as husbanding the limited capacity of the system to the greatest good of all the users and applications.

The second task, providing a consistent application interface, is especially important if there is to be more than one of a particular type of computer using the operating system, or if the hardware making up the computer is ever open to change. A consistent application program interface (API) allows a software developer to write an application on one computer and have a high level of confidence that it will run on another computer of the same type, even if the amount of memory or the quantity of storage is different on the two machines.

Even if a particular computer is unique, an operating system can ensure that applications continue to run when hardware upgrades and updates occur. This is because the operating system and not the application are charged with managing the hardware and the distribution of its resources.

KINDS OF OPERATING SYSTEM

There are generally four types of operating system which are categorized on the basis of types of computers they control and the sort of applications they support. The categories are:

Real-time operating system (RTOS) : Real-time operating systems are used to control machinery, scientific instruments and industrial systems. An RTOS typically has very little user-interface capability, and no end-user utilities, since the system will be a "sealed box" when delivered for use. A very important part of an RTOS is managing the resources of the computer so that a particular operation executes in precisely the same amount of time every time it occurs.

Single-user, single task : According to its name, this operating system is designed to manage the computer so that one user can effectively do one thing at a time. The *Palm OS* for

Palm handheld computers is an example of a modern single-user, single-task operating system.

Single-user, multi-tasking : This type of operating system is used by most people on their desktop and laptop computers today. Microsoft's *Windows* and Apple's *MacOS* platforms are both examples of operating systems that will let a single user have several programs in operation at the same time. For example, it's entirely possible for a Windows user to be writing a note in a word processor while downloading a file from the Internet while printing the text of an e-mail message.

Multi-user : A multi-user operating system allows many different users to take advantage of the computer's resources simultaneously. The operating system must make sure that the requirements of the various users are balanced, and that each of the programs they are using has sufficient and separate resources so that a problem with one user doesn't affect the entire community of users. UNIX, VMS and mainframe operating systems, such as *MVS*, are examples of multi-user operating systems.

It's important to differentiate here between multi-user operating systems and single-user operating systems that support networking. *Windows 2000* and *Novell Netware* can each support hundreds or thousands of networked users, but the operating systems themselves aren't true multi-user operating systems. The network support and all of the remote user logins the network enables are, in the overall plan of the operating system, a program being run by the administrative user.

MEMORY STORAGE AND MANAGEMENT BY OPERATING SYSTEM

When an operating system manages the computer's memory, there are two broad tasks to be accomplished:

1. Each process must have enough memory in which to execute, and it can neither run into the memory space of another process nor be run into by another process.

2. The different types of memory in the system must be used properly so that each process can run most effectively.

The first task requires the operating system to set up memory boundaries for types of software and for individual applications.

When applications begin to be loaded into memory, they are loaded in block sizes determined by the operating system. If the block size is 2 kilobytes, then every process that is loaded will be given a chunk of memory that is a multiple of 2 kilobytes in size. Applications will be loaded in these fixed block sizes, with the blocks starting and ending on boundaries established by words of 4 or 8 bytes. These blocks and boundaries help to ensure that applications won't be loaded on top of one another's space by a poorly calculated bit or two. With that ensured, the larger question is what to do when the 500-kilobyte application space is filled.

In most computers, it's possible to add memory beyond the original capacity. One might expand RAM from 1 to 2 megabytes. This works fine, but tends to be relatively expensive. It also ignores a fundamental fact of computing — most of the information that an application stores in memory is not being used at any given moment. A processor can only access memory one location at a time, so the vast majority of RAM is unused at any moment. Since disk space is cheap compared to RAM, then moving information in RAM to hard disk can greatly expand RAM space at no cost. This technique is called virtual memory management.

Disk storage is only one of the memory types that must be managed by the operating system, and is the slowest. Ranked in order of speed, the types of memory in a computer system are:

High-speed cache : This is fast, relatively small amounts of memory that are available to the CPU through the fastest connections. Cache controllers predict which pieces of data the CPU will need next and pull it from main memory into high-speed cache to speed up system performance.

Main memory : This is the RAM that measured in megabytes when.

Secondary memory : This is most often some sort of rotating magnetic storage that keeps applications and data available to be used, and serves as virtual RAM under the control of the operating system.

The operating system must balance the needs of the various processes with the availability of the different types of memory, moving data in blocks (called pages) between available memories as the schedule of processes dictates.

DEVICE MANAGEMENT BY OPERATING SYSTEM

The path between the operating system and virtually all hardware on the computer's motherboard goes through a special program called a driver. Driver's function is of the translator between the electrical signals of the hardware subsystems and the high-level programming languages of the operating system and application programs. Drivers take data that the operating system has defined as a file and translate them into streams of bits placed in specific locations on storage devices, or a series of laser pulses in a printer.

There are such wide differences in the hardware controlled through drivers, there are differences in the way that the driver programs function, but most are run when the device is required, and function much the same as any other process. The operating system will frequently assign high-priority blocks to drivers so that the hardware resource can be released and readied for further use as quickly as possible.

One reason that drivers are separate from the operating system is so that new functions can be added to the driver — and thus to the hardware subsystems — without requiring the operating system itself to be modified, recompiled and redistributed. Through the development of new hardware device drivers, development often performed or paid for by the manufacturer of the subsystems rather than the publisher of the operating system, input/output capabilities of the overall system can be greatly enhanced.

Managing all the resources of the computer system is a large part of the operating system's function and, in the case of real-time operating systems, may be virtually all the functionality required. For other operating systems, though, providing a relatively simple, consistent way for applications and humans to use the power of the hardware is a crucial part of their reason for existing.

OPEN SOURCE OPERATING SOFTWARE

An important question concerning the future of operating systems revolves around the ability of a particular philosophy of software distribution to create an operating system useable by corporations and consumers together. Linux, the operating system was created and distributed according to the principles of **open source**, has had a significant impact on the operating system in general. Most operating systems, drivers and utility programs are written by commercial organizations that distribute **executable versions** of their software — these versions can't be studied or altered. Open source requires the distribution of original source materials that can be studied, altered and built upon, with the results once again freely distributed. In the desktop computer realm, this has led to the development and distribution of countless useful and cost-free applications like the image manipulation program (GIMP) and the popular web server Apache.

BIOS (BASIC INPUT OUTPUT SYSTEM)

This is essential code that initializes the hardware in PC and loads the OS. It can loosely be considered to be a part of the OS. When a PC is turned on it is under the control of BIOS code which is stored in a special memory chip on the mother board. The BIOS store configuration information about all the hardware in the PC. The type of hard drive, floppy drive, video display, CD etc. is kept here so that the PC knows how to communicate with these devices. Newer BIOS code has 'auto' modes that can figure out what hardware is attached and how to communicate with it. In older BIOS, if hardware is changed without updating the BIOS the PC may not complete it's boot up cycle, and will just sit there until the new device parameters are loaded into it.

After BIOS copies the OS code from hard drive into memory, the copy of the OS in memory takes over and finished the initialization process. When it is complete the PC is ready to use application programs.

APPLICATION PROGRAMS

These are called workhorses of code. They are the most familiar software. They run on top of the OS and allow user to use PC for work or play. Word, Excel, Frame Maker, Corel Draw and PC games are a just a few examples of the thousands of Application Programs that are available.

Many programs come with (bundled) the Windows Operating System. The following applications represent a tiny fraction of the programs. The list of available application software is immense. Software of every type is available to do almost anything one needs or wants to do.

Applications found in Start > Programs

* Windows Explorer - It allows seeing and accessing all the folders and files on HDD, FDD, CDD and Local Area Network.
* Internet Explorer - This is a browser program which enables PC receive web pages from servers located anywhere on the Internet.
* MS-DOS Prompt - It allows to open a DOS session under Windows.

Applications found in Start > Programs > Accessories

* Notepad - Simple word processor with limited features used to create and modify text (.txt) files.

* Wordpad - Word processor with more features than Notebook. It can handle large document (.doc) files.
* Paint - Allows to create, modify and save image files in bitmap (.bmp), JPEG (.jpg), GIF (.gif) formats.
* Imaging - Another image file program that works with bitmap (.bmp), and TIFF (.tif) filetypes.
* Sound recorder - If a microphone is plugged into sound card one can record sounds and play them back with option of adding effects like echo. One can also play them backwards as well.
* Windows Media Player - Plays video (.mpg) and audio (.wav) files.
* Calculator - Simple calculator with half a dozen functions.

Other Application Programs

In addition to bundled applications, there are thousands of third party application programs one can use. Each program specifies the minimum required free hard drive capacity, processor speed, video resolution and amount of installed memory in order for it to operate properly.

Microsoft Office

MS Office has become a de facto standard for creating and sharing documents for a large percentage of both business and personal PC users. It includes:

* WORD is a powerful word processor that contains templates that allow to quickly creating numerous document types such as letters, faxes and resumes. Tables for chart type data can quickly be set up.

It supports a wide variety of fonts, font sizes, and text foreground and background colors.

One can format it so that envelopes and labels can be printed.

* EXCEL is a spread sheet program for creating and manipulating numerical data. It contains functions that allow powerful calculations for general, accounting and engineering

type problems. Charts and graphs can be generated to better show relationships between variables.

* POWERPOINT enables to create powerful slide show presentations. Color, animation and a number of effects can be used to enhance slide show.

* ACCESS is a database program that can be used to keep track of large amounts of related data that can be output in any manner.

APPLICATION SOFTWARE

Application software is a subclass of computer software that employs the capabilities of a computer directly to a task that the user to perform. This is contrasted with system software which is involved in integrating a computer's various capabilities, but typically does not directly apply them in the performance of tasks that benefit the user. The term application refers to both the *application software* and its implementation.

The exact delineation between the operating system and application software is not precise. Typical examples of software applications are word processors, spreadsheets, and media players.

Application suite

Multiple applications bundled together as a package are referred to as an application suite. Microsoft Office and OpenOffice.org, which bundle together a word processor, a spreadsheet, and several other discrete applications, are typical examples. The separate applications in a suite usually have a user interface that has some commonality making it easier for the user to learn and use each application. For example a spreadsheet might be able to be embedded in a word processor document even though it had been created in the separate spreadsheet application.

User-written software

User-written software tailors systems to meet the user's specific needs. User-written software include spreadsheet templates, word processor macros, scientific simulations, graphics and animation scripts. Even email filters are a kind of user software.

Application software classification

There are many types of Application Software:

* Enterprise Software addresses the needs of organization processes and data flow, often in a large distributed ecosystem. Examples include Financial, Customer Relationship Management, and Supply Chain Management.
* Enterprise Infrastructure Software provides common capabilities needed to create Enterprise Software systems. Examples include Database, Email servers, and Network and Security Management.
* Information Worker Software addresses the needs of individuals to **create** and manage information, often in organizational settings. Examples include Analytical, Collaborative, Word Processing, Spreadsheets, Email and Blog clients, Personal Information Systems, and Individual Media editing.
* Media and Entertainment Software addresses the needs of individuals and groups to consume digital entertainment and **published** digital content. Examples include Media Players, Web Browsers, Help browsers, and Games.
* Educational Software is related to Media and Entertainment Software, but has distinct requirements for delivering evaluations (tests) and tracking progress through material. It is also related to collaboration software in that many Educational Software systems include collaborative capabilities.
* Media and Software Development Software addresses the needs of individuals who generate software or media for others to consume, most often in a commercial or

educational setting. This type includes Integrated Development Environments, Application Programmer Interfaces, HTML editors, Digital Animation editors, Digital Audio and Video composition, and many others.

ENTERPRISE SOFTWARE

* Accounting software
* Business management systems
* Contracts and Agreements Management
* Customer relationship management
* Enterprise Records Management
* Electronic business software
* Enterprise resource planning
* Human Resource Management
* Legal Matter Management
* Management information systems
* Operational risk management
* Product Lifecycle Management
* Project management software
* Supply chain management

Information Worker Software

* Time and Resource Management
  — Task and Scheduling
* Data Management
  — Contact Management
  — Spreadsheet
  — Personal Database
* Documentation
  — Word Processing
  — Desktop publishing software
  — Diagramming Software
  — Presentation software
* Analytical software

PRODUCT ENGINEERING SOFTWARE

Educational Software

* Classroom Management
* Survey Management
* Training Management

Media Development Software

* Graphic art software

DIFFERENT OTHER SOFTWARE

Retail software : This type of software is sold off the shelves of retail stores. It includes expensive packaging designed to catch the eye of shoppers and, as such, is generally more expensive. An advantage of retail software is that it comes with printed manuals and installation instructions, missing in hard-copy form from virtually every other category of software. However, when hard-copy manuals and instructions are not required, a downloadable version off the Internet will be less expensive, if available.

OEM software : OEM stands for "Original Equipment Manufacturer" and refers to software sold in bulk to resellers, designed to be bundled with hardware. For example, Microsoft has contracts with various companies including Dell Computers, Toshiba, Gateway and others. Microsoft sells its operating systems as OEM software at a reduced price, minus retail packaging, manuals and installation instructions. Resellers install the operating system before systems are sold and the OEM CD is supplied to the buyer. The "manual" consists of the Help menu built into the software itself. OEM software is not legal to buy unbundled from its original hardware system.

Shareware : This kind of software are downloadable from the Internet. Licenses differ, but commonly the user is allowed to try the program for free, for a period stipulated in the license, usually thirty days. At the end of the trial period, the software must be purchased or uninstalled. Some shareware incorporates an internal clock that disables the program after the trial period unless a serial number is supplied.

Crippleware : This software is similar to shareware except that key features will cease to work after the trial period has ended. For example, the "save" function, the print function, or some other vital feature necessary to use the program effectively may become unusable. This "cripples" the program. Other types of crippleware incorporate crippled functions throughout the trial period. A purchase is necessary to unlock the crippled features.

Demo software : Demo software is not intended to be a functioning program, it only allows partial functioning. It is mainly designed to *demonstrate* what a purchased version is capable of doing, and often works more like an automated tutorial. If a person wants to use the program, they must buy a fully functioning version.

Adware : This is free software that is supported by advertisements built into the program itself. Some adware requires a live Internet feed and uses constant bandwidth to upload new advertisements. The user must view these ads in the interface of the program. Disabling the ads is against the license agreement. Adware is not particularly popular.

Spyware : Spyware software is normally free, but can be shareware. The concept of spyware is very unpopular, and many programs that use spyware protocols were forced to disclose this to users and offer a means to turn off reporting functions. Other spyware programs divulge the protocols in their licenses, and make acceptance of the spyware feature a condition of agreement for using the software.

Freeware : Freeware is also downloadable off the Internet and free of charge. Often freeware is only free for personal use, while commercial use requires a paid license. Freeware does not contain spyware or adware. If it is found to contain either of these, it is reclassified as such.

Public domain software : This is free software, but unlike freeware, public domain software does not have a specific copyright owner or license restrictions. It is the only software that can be legally modified by the user for his or her own purposes.

SYSTEMS SOFTWARE

Systems software is a set of programs that organise, utilise and control hardware in a computer system. It also refers to the operating system and all utility programs that manage computer resources at a low level. Systems software includes compilers, loaders, linkers, and debuggers. System software is a generic term referring to any computer software which manages and controls the hardware so that application software can perform a task. It is an essential part of the computer system. An operating system is an obvious example, while an OpenGL or database library are less obvious examples. System software contrasts with application software, which are programs that help the end-user to perform specific, productive tasks, such as word processing or image manipulation. If system software is stored on non-volatile storage such as integrated circuits, it is usually termed firmware.

a. Compilers

It is a program that translates *source code* into *object code.* The compiler derives its name from the way it works, looking at the entire piece of source code and collecting and reorganizing the instructions. Thus, a compiler differs from an *interpreter,* which analyzes and executes each line of source code in succession, without looking at the entire program. The advantage of interpreters is that they can execute a program immediately. Compilers require some time before an executable program emerges. However, programs produced by compilers run much faster than the same programs executed by an interpreter. Every high-level programming language (except strictly interpretive languages) comes with a compiler. In effect, the compiler is the language, because it defines which instructions are acceptable. Because compilers translate source code into object code, which is unique for each type of computer, many compilers are available for the same language. For example, there is a FORTRAN compiler for PCs and another for Apple Macintosh computers. In addition, the compiler industry is quite competitive, so there are actually many compilers for each language on each type of computer. More than a dozen companies develop and sell C compilers for the PC.

b. Loaders

It is an operating system utility that copies programs from a storage device to main memory, where they can be executed. In addition to copying a program into main memory, the loader can also replace virtual addresses with physical addresses. Most loaders are transparent, i.e., one cannot directly execute them, but the operating system uses them when necessary.

c. Linker

It is also called *link editor* and *binder,* a linker is a program that combines object modules to form an executable program. Many programming languages allow to write different pieces of code, called *modules*, separately. This simplifies the programming task because one can break a large program into small, more manageable pieces. Eventually, though, user needs to put all the modules together. This is the job of the linker. In addition to combining modules, a linker also replaces symbolic addresses with real addresses. Therefore, one may need to link a program even if it contains only one module.

d. Debugger

It is a special program used to find errors (*bugs*) in other programs. A debugger allows a programmer to stop a program at any point and examine and change the values of variables

SOME COMMERCIAL TYPES OF SOFTWARE

Many commercial software are available. In financial software packages various types of software are e-commerce software, bar code software, inventory software, billing software applications for medical offices, mortgage, real estate and loan servicing software, retail software, as well as stock trading applications. Artists will find valuable information in the music software and photo/video editing software.

Some are parental control software, educational software, childcare software, and daycare software. Other interesting software include free D burning software, home design software, embroidery software, and even such highly specialized applications as trucking software, landscape software and astrology software.

Examples of various types of software are as following:

1. Accounting software
2. Ecommerce software
3. Free CD burning software
4. Video editing software
5. Computer software
6. Inventory software
7. Educational software
8. Medical billing software
9. Mortgage software
10. Home design software
11. Bar code software
12. Parental control software
13. Real estate software
14. Embroidery software
15. Retail software
16. Photo editing software
17. Loan servicing software
18. Music software
19. Trucking software

20. Landscape software
21. Day care software
22. Stock trading software
23. Astrology software
24. Childcare software

FREE WARE FOR LIBRARY AND INFORMATION MANAGEMENT

1. Automated Classification and Thesaurus Construction

Software Title	*Operating Systems*	*Description*
ExtPhr32	Windows 32	Extracts every word and every phrase up to a certain length that occurs at least a minimum number of times in a source text and that does not start or end with a stopword.
NEPHIS32	Windows 32	Incorporates the classic NEPHIS string indexing system, with some additional formatting features, including production of indexes in HTML format.
TexNet32	Windows 32	Assists in the writing of abstracts and other short summaries. Includes word and phrase extraction and various other capabilities.
TheW32	Windows 32	Assists in creating, modifying, and printing out a thesaurus. Allows users to define their own link types and report formats. Includes a template with definitions of the standard SN, USE, UF, BT, NT, and RT relation types and alphabetical, short, and tree report formats.
XRefHT3	Windows 32	Assists in human indexing of Web documents. Produces index displays in HTML form, including links to Web documents. Allows for automatic target extraction, target and link insertion, and production of tables of contents in HTML files and insertion of certain other HTML codes in plain text files.

2. Knowledge Management

Software Title	*Operating Systems*	*Description*
Apollo CMS	Unix like systems	Apollo CMS is a content management system written in Python, using PostgreSQL as a backend. It is written in a modular way allowing for new types of content, and new database backends
Apache:: PageKit	Unix like systems	PageKit is a modular application framework that uses HTML::Template and XML to separate the design from the content. It includes session management, authentication, form validation, co-branding, and a content management system.
SPIN CMS	Unix like systems	SPIN CMS is a basic and simple content management system for newbies and those wanting something up and running quickly. It can also store images and other binary files. The admin mode allows one to add, view, edit, and delete content arranged in multiple categories. Various output options are available for 'public' pages. Based on MySQL and PHP 4.
Baseit	Unix like systems	basit is a Web site content management system developed using PHP. The package provides a Web-based user interface to configure all the attributes of the Web site and add/edit/delete articles. A full usage manual including setup, administration, and maintenance is provided. The package doesn't need any database support on the Web server.
BaseML	Unix like systems	BaseML is a tool/compiler to manage and construct Web sites. To create a Web site, library provide BaseML with a set of "source" XML files that contain the styling and contents of the web site. The source code format is a mix between XML/XHTML components and procedural programming.

AStA-CMS	Unix like systems	AStA-CMS is a simple content managment system (CMS) which allows library to manage the contents of the Web site in a well-structured manner, based on an hierarchieecal category tree. It also allows library to store files and images in a database. It features multi-user functionality and the ability to manage articles, files, and images easily. It is based on templates, and thus provides a proper separation of PHP code and visual design.
asCMS		asCMS is an easy-to-use, powerful PHP-based content management system.
Ascend	Unix, Windows, MacOS	Ascend is an advanced profesional Web content manager. It supports multiple OSes, multiple database connections, multiple languages, multiple presentation schemes, multiple browsers, XLM integration, a Java editor, advanced security management for both users and administrators, integrated CRM, and CVS integration. It can also use CGI scripts or ascend-specific plugins to extend the software for tasks such as e-commerce, portals, intranets, Unix system administration, and editorial systems.
Arsdigita CMS	OS Independent	Arsdigita CMS is a powerful content management system. It has a task list for production staff to track their assignments and the status of their current work items; a site map browser to view and organize pages and content items and determine access control to branches of the site; a standard interface for creating, editing, approving, and deploying content items; a template manager for creating, editing, and organizing presentation templates and related assets; a metadata manager for viewing and defining content types and associations; a category browser for

		managing a hierarchy of subject headings that may be applied to content items; and administrative and management tools for creating and editing user attributes and tracking global workflow statistics.

3. Database Systems

Software Title	*Operating Systems*	*Description*
XBase	GNU Linux, Windows 98, Windows 95, Windows 2000	Xbase DBMS is a collection of specifications, programs, utilities and a C++ class library for manipulating Xbase type datafiles and indices which has been developed and produced by many people from around the world using the internet. This project was formerly known as Xbase for Linux, but several inputs are received that people are compiling this on platforms other than Linux, it is have renamed to Xbase DBMS. The main development of this library however, remains on the Linux platform utilizing the GCC public domain C/C++ compiler.
Haccess 97	Windows	A reverse engineer the Microsoft access (MDB / jet) database format, and provide full documentation and cross platform ansi C++ read-write libs to support SQL, Cursors, ODBC drivers, GUI's and data-recovery.
genSQL	GNU Linux, Unix	genSQL is a generic SQL database UI, written in C++ using the GNOME—/ GTK— widget sets. Its aim is to provide a complete and easy-to-use interface to various SQL databases.To create a generic SQL GUI for Linux/UNIX platforms.Opening and closing multiple databases with different types, Viewing the list of tables in the databases, Dropping tables, Viewing table data

		Loading and unloading GQL database drivers.
MDBMS	GNU Linux	MDBMS is a new relational database management system. It is very clean and small, yet powerful enough to do complex database work. MDBMS is currently available in source and binary form, for Linux and BSD systems. MDBMS is known to work correctly on Alpha-linux and Sun Solaris machines.
SQLite	Unix-like systems, Windows	SQLite is a C library that implements an SQL database engine. A large subset of SQL92 is supported. A complete database is stored in a single disk file. The API is designed for convenience and ease of use. Applications that link against SQLite can enjoy the power and flexiblity of an SQL database without the administrative hassles of supporting a separate database server. Because it omits the client-server interaction overhead and writes directly to disk, SQLite is also faster than the big database servers for most operations. In addition to the C library, the SQLite distribution includes a command-line tool for interacting with SQLite databases and SQLite bindings for Tcl/Tk.
Mimer SQL	GNU Linux, Unix, Windows NT, Windows 98, Windows 95, Windows 2000	Mimer SQL is an SQL-based, high-performance RDBMS, targeted at mission-critical client/server environments. It offers scalable performance (including multi-processor support), and is available in all major Unix (including Linux), Windows, and OpenVMS platforms. It offers Optimistic Concurrency Control as the transaction handling method, instead of the more commonly used locking-type.
WebMySQL	GNU Linux, Unix	WebMySQL is a Web interface to a MySQL database server. It supports the following SQL queries: SELECT,

		DESCRIBE, JOIN, and SHOW, and advanced query options such as LIMIT, DISTINCT, and GROUP. It has a fully customisable look and feel, and outputs either HTML results or a downloadable CSV file.
JCDatabase	GNU Linux, Unix, Windows NT, Windows 98, Windows 2000	JCDatabase, Direct File Database Management System has the following advantages over JDBC: 1. JCDatabase does not require library to run a database server or configure the system to provide access to a local database and; 2. JCDatabase can be very conveniently be deployed as part of a traditional client-side application using direct file access.
Maccess	GNU Linux	Maccess helps to analyse data in databases; it is a Web interface to an SQL server. Its main features are user friendly query creation for SQL server; selective, grouped, and crosstab queries; various outputs (XML, HTML, TXT, graphs, and Europe map); data ungrouping in grouped queries; creation of Web links and definition of CSS styles for all values in output; ability to format numbers, dates, and times; and input and data update.
GigaBase	GNU Linux, Unix, Windows NT, FreeBSD, OS/2, Windows 98, Solaris, MAC OS, AIX	GigaBASE is object-relational embedded database engine for C++ applications. It provides SQL-like query language, smart C++ interface (loading objects instead of tupples), transaction based on shadowing page algorithm (no separate log file and very fast)
Dbreport	GNU Linux, Unix	dbreport creates a PostScript report of a table or view within a database. The database is accessed via unixODBC, and the format for the report is specified by an XML file.
NetSQL	GNU Linux, Solaris	NetSQL is a light and robust SQL database server for Linux and Sun

		Solaris. Because it can be configured on a virtual web server, it enables every Webmaster with a virtual web site to host database applications. It comes with a very intuitive web application builder, DataDiplomat, which not only has a visual form editor, CGI script generator and installer, but also provides necessary ODBC database support. With DataDiplomat application builder, library can create searchable web sites with real-time updating capability without programming.
ACDK SQL	Windows NT, Solaris, Windows, Windows 2000	ACDK is a platform-independent library for generating distributed server-based components and applications. acdk_sql contains the driver model for accessing SQL databases, which is quite similar to the JDBC interface.
Bonddb	Unix like systems	Bonddb is a object oriented wrapper for postgresql SQL. It's a fast data abstraction layer written in C for C/C++ applications to allow easy access to class objects. Bonddb is not a stand alone object orientated database, but works on top of SQL so library still get a solid proven backend with the benefits of things like prescient objects.
GOODS	Windows	GOODS is an object-oriented fully distributed database management system using an active client model. Its multithreaded high-performance database server is language and application independent. The client application interface is built using meta-object protocol and provides transparent persistency for normal programming languages, and the aspect-oriented programming approach used for implementing object access and synchronization policies makes it possible to define various database access strategies fitting requirements of

		concrete application and separate them from application code itself.
KDBMS	Unix-like systems	KDBMS is a database management system with its own interpreted scripting language supporting WEB/WAP/IVR. It also supports multiprocessing, shared memory, and many other things.
db4o (Database for Objects)	Platform Independent	db4o (database for objects) is a client/server Java object database. It features transactions, automatic class recognition, query-by-example, callback methods, memory files, cascade-on-delete, multi-threaded access, and Java Webstart and servlet support.
UML2SQL	Platform Independent	UML2SQL is a Java tool which synchronizes SQL databases with UML diagrams. UML class diagrams can thus serve as authoritative definitions of database catalogs. UML2SQL can be used together with a UML design tool which is capable of producing XMI. For the most generic operations, only a JDBC driver is required for the SQL database. Vendor specific DDL commands for unsupported databases can be easily implemented by extending the DatabaseAdapter class.
Orac DB Tool	Platform Independent	Orac is a GUI tool for Database Administrators written in Perl, Tk, and DBI. It organizes many useful DBA scripts into a nice graphical interface, for collecting statistics on the database, or diagnosing problems. It currently supports Oracle, Sybase, and Informix. Richard Sutherland\
FutureSQL Tool	GNU Linux, Unix	FutureSQL is a Rapid Application Development web database administration tool written in Perl. FutureSQL allows one to easily setup config files to view, edit, delete and otherwise process records from a MySQL database. It uses a data

		dictionary, configuration files and html templates, and allows \"pre-processing\" and \"post-processing\" on both the fields, records and operations. It allows multiple views and operations on a data set, including the use of joined tables for queries and reports. A demo application with most of the features is included.
Qddb	GNU Linux, Unix, FreeBSD, Solaris, HPUX, Sun OS, Open BSD	Qddb is powerful database creation software. Using Qddb, library can quickly and easily create database applications. nxqddb allows library to create a complete database application in less than five minutes, and Hx lets library manage database information over the web through a CGI interface. Larger applications can also be developed using Fx, a powerful TCL/Tk toolkit.
MaVerick - MultiValue Database Management System	Platform Independent	MaVerick is a MultiValue database management system.MultiValue databases can be relational, but their method of storing data is much more flexible than the 2 dimensional table library normally associate with relational databases, i.e., library can store multiple fields within a field. On top of which, they are extremely fast in operation.
XDBM	Windows	XDBM is a database designed specifically to handle XML data. It offers greater speed (since the data is stored in a preparsed format), lower memory usage (since only those parts of the file needed are loaded into memory) and faster searching for XML elements. XDBM can also parse and store plain text XML files. XDBM aims to be as close to the W3\
LocalSQL	GNU Linux, Windows	LocalSQL provides a library and command line tool for direct invocation and processing of SQL statements on

		local databases. A subset of SQL92 is implemented, notably 'INSERT UPDATE DELETE SELECT CREATE TABLE CREATE INDEX DROP TABLE'.
RedBase Pure Java RDBMS	Platform Independent	RedBase Pure Java RDBMS is a 100% Pure Java database, with an ultra-compact footprint designed for rapidly developing applications that extend enterprise data management capabilities, to mobile and embedded devices. It is ideal for mobile, wireless, and embedded applications, and it delivers essential relational database functionality in a small footprint, while providing flexible data access, and the familiar feel through entry SQL-92 compliance, and JDBC access.
Scientific Image Database (SIDB)	GNU Linux	SIDB (Scientific Image Database) is a Web-driven database for (scientific) images. Entry of image meta-data is facilitated through the use of user-definable templates. Users have complete control over who else may see the submitted image and data. SIDB offers various views on the actual image data. Thumbnail plus meta-data can be printed, and multiple thumbnails can be combined in galleries. A freeware version of the Huygens software (http://www.svi.nl) can be used to generate projections from 3-D images (most confocal microscope image formats are supported), as well as MPEG movies, showing the individual layers from the 3-D images.
Ozone Java ODBMS	Platform Independent	ozone is a transaction-based, object-oriented database management system completely implemented in Java and distributed under an open-source license. It aims to evolve a technology that allows developers to build pure object-oriented, Java database

		applications. ozone also contains a full-featured XML API that allows library to use it as an XML repository.
Lago Database		GNU Linux, FreeBSD, Windows Lago is a portable (Linux/Windows), multi-threaded database written in C++.

4. Indexers, Information Retrieval and Digital Library Packages

Software Title	*Operating Systems*	*Description*
FreeWAIS-sf	Unix-like systems	FreeWAIS is a text retrieval system based on the vector space model. freeWAIS-sf is an extension of freeWAIS, developed at the computer science department of the University of Dortmund. sf means structured fields and indicates the main improvement of freeWAIS-sf.The SF suffix in the software name stands for structured fields an indexing and search feature which distinguishes this software from its predecessors.
Greenstone Digital Library Software:	Unix, Windows NT, Windows 98, Windows 95	Greenstone is a suite of software which has the ability to serve digital library collections and build new collections. It provides a new way of organizing information and publishing it on the Internet or on CD-ROM. Greenstone is produced by the New Zealand Digital Library Project at the University of Waikato, and distributed in cooperation with UNESCO and the Humanities Library Project.
Harvest Web Indexer:	Unix like Systems	The Harvest indexer offers a distributed solution to the problems of indexing data made available on the web. With each web server running a local Gatherer feeding into a central Broker many of the problems of web crawling are avoided. The Harvest Indexer can fetch and index data made available by

		HTTP, Gopher, FTP, or NNTP. It has summarisers capable of indexing data in a wide variety of file formats.
ht://Dig	GNU Linux, Solaris, HPUX, Sun OS, IRIX	http://Dig system is a complete indexing and searching system for a domain or intranet. It has the ability to search through many servers on a network by acting as a WWW browser.
Perlfect	GNU Linux, Unix-like systems, Windows	Perlfect Search is a customizable and effective site indexing/searching suite. It comes as a pair of disctinct scripts. The indexer, that automatically scans and indexes a web site, and the search engine, a cgi script that serves search queries for keywords over the index, and displays results pages in html, in a standard format including relevance ranking for each matching document.
Swish-E	Unix-like systems, DOS, Windows NT	The Simple Web Indexing System for Humans-Enhanced is a system for indexing collections of Web pages or other text files. The AND, OR, and NOT operators are supported, words can be truncated using *, and searches can be limited to particular fields or to certain HTML tags. Index files consist of only one file, so they are portable and can be easily maintained.
Isearch	GNU Linux, Unix, FreeBSD, Solaris, HPUX, IRIX, Ultrix	Isearch is an open source software package for indexing and searching text documents. It supports full text and field based searching, relevance ranked results, and Boolean queries, and it can index many kinds of documents, including HTML, mail folders, list digests, and text with SGML-style mark up. Additional document types can be added as C++ modules. Isearch also includes CGI tools to provide a web interface for searching.
mno GoSearch	GNU Linux, Unix, Unix-like systems, FreeBSD, Solaris,	Formerly known as UdmSearch, mnoGoSearch is a full-featured SQL based web search engine. It has number

	HPUX, Sun OS, Open BSD, Windows	of unique features, which make it capable of wide range of application - from search within the site to a specialized search system.
Webinator	GNU Linux, Windows NT, FreeBSD, Solaris, HPUX, IRIX, AIX, Windows 2000, DEC UNIX, Tru64	The Webinator is a sophisticated Web index and retrieval package. It allows the Website administrator to create and provide a high quality retrieval interface to collections of web documents
Zebra	Unix-like systems, Windows	Zebra is a fielded free-text indexing and retrieval engine with a Z39.50 frontend. Library can use any compatible, commercial or freeware Z39.50 client to access data stored in Zebra.
Amberfish	Unix like systems	Amberfish™ is Etymon's new high performance text retrieval software.
Locus	Linux	Locus is a fulltext database search engine
Cheshire II	Unix like systems	The Cheshire II project is developing a next-generation online catalog and full-text information retrieval system using advanced IR techniques. This system is being deployed in a working library environment and its use and acceptance by local library patrons and remote network users are being evaluated. The Cheshire II system was designed to overcome twin problems of topical searching in online catalogs, search failure and information overload as well as to provide a bridge between the purely bibliographic realm of previous generations of online catalogs and the rapidly expanding realm of full-text and multimedia information resources. The system incorporates a client/server architecture with implementations of current information retrieval standards including Z39.50 and SGML and XML.
OCLC Site Search		Windows NT Server, Unix like systems The OCLC SiteSearch™ software provides a comprehensive solution for

		managing distributed library information resources in a World Wide Web environment. It offers tools that integrate electronic resources under one web interface, provide flexible access to resources, and build text and image databases locally.
Ganesha Digital Library	Windows 98, NT, 2000	Features: Distributed Knowledge Management, Centralized Knowledge Distributiont, Online Member Registration, Roaming Membership, Searching, Category-Based Organization, Upload Metadata and Files, Personal Directories, Review Forum, Access Restriction, Image Thumbnail, Knowledge Organization, News, Synchronization, Member and Group Administration, Statistics, Advertisement, Dublin Core / IndonesiaDLN Metadata, XML Based Transaction, CD-ROM Enabled

5. Library Automation

Software Title	*Operating Systems*	*Description*
PhpMy Library	Linux	PhpMyLibrary is a PHP MySQL Library automation application. The program consist of cataloging, circulation, and the webpac module. The programs also has an import export feature. The program strictly follow the USMARC standard for adding materials.
KOHA		Claiming to be the first Free Open Source Library System

6. Miscellaneous

Software Title	*Operating Systems*	*Description*
Concordance	Unix like Sytems	Concordance is a program that scans a text file and outputs concordance lines

		based on a node entered by the user. There are two web-based versions of Concordance: one in PERL and the other in C.
Rakim (Rakim A Knowledge Instant Messanger)	Unix like Systems	rakim is an attempt at providing a web-based reference service. It is very similar to standard chat room scripts, but with some added features that might be useful when trying to help patrons find answers to their questions.
WIBS	Unix like systems	WIBS stands for Windsor Internet Booking System. It is based on the wonderful MRBS (Meeting Room Booking System) and is designed to mediate access to public stations in libraries. WIBS features include: - uses the browser for logon and time -control -is platform-independent -allows patrons to book stations on the web
Reference Desk Manager (RDM)	Unix-like System	RDM features are: Email weblog — with search feature Electronic Card File — with search feature Common Links Area Web-based Administration
ROADS	Unix-like System	ROADS is a set of software tools to enable the set up and maintenance of Web based subject gateways.

REFERENCES

1. chortle.ccsu.edu/CS151/cs151java.html
2. Ashley, N.W. (1995) *Diffusion of network information retrieval in academia*. Unpublished doctoral dissertation, The University of Arizona, Tucson.
3. Baker, D. (1978) "Characteristics of the literature used by English musicologists." *Journal of Librarianship*, 10, 183-200.
4. Baker, P.R. (1985) "Adoption of a computer software cataloging system at an elementary school." *School Library Media Quarterly*, 13(3/4), 208-214.
5. www.compinfo-center.com
6. www.omsar.gov.lb/ICTSG/202SW/2.1_The_Scope_of_Software_Applications.htm

7. www.grassrootsdesign.com/intro
8. www.businesslink.gov.uk/bdotg/action/layer?topicId=1073935242
9. www.jegsworks.com/Lessons/lesson1-2/lesson1-1.htm
10. www.onestoptesting.com/introduction
11. www.newfreedownloads.com/Information-Management/index.html
12. www.freedownloadscenter.com/Information_Management/free/alpha25.html
13. Bates, M.J., Wilde, D.N., and Siegfried, S. (1993) "An analysis of search terminology used by humanities scholars: The Getty Online Searching Project Report Number 1." *Library Quarterly*, 63(1), 1-39.
14. Bebout, L., Davis, D., Jr., and Oehlerts, D. (1975) "User studies in the humanities: A survey and a proposal." *RQ*, 15, 40-44.
15. Cohen, J. (1996) "Computer mediated communication and publication productivity among faculty." *Internet Research*, 6(2/3), 41-63.
16. www.britannica.com/eb/article-218049/information-system
17. seamonkey.ed.asu.edu/emc300/software/types.html
18. www.bls.gov/oco/ocos110.htm
19. Chatman, E.A. (1986) "Diffusion theory: A review and test of a conceptual model in information diffusion." *Journal of the American Society for Information Sciencc*, 37(6), 377-386.
20. Clegg, S.M. (1985) "User surveys and statistics-- The opportunities for music libraries." *Fontes Artis Musicae*, 32(1), 69-75.
21. www.webopedia.com/TERM/S/software.html
22. www.webopedia.com/TERM/A/systems_software.html
23. techweb.com/encyclopedia/defineterm.jhtml
24. herpolhode.com/rob/utah2000.pdf
25. Berg, B.L. (1995) *Qualitative research methods for the social sciences*. 2nd ed. Boston: Allyn and Bacon.
26. Berman, Y. (1996) "Discussion groups on the Internet as sources of information: The case of social work." *Aslib Proceedings*, 48(2), 31-36.
27. Bridges, A.E. and Clement, R.T. (1997) "Crossing the threshold of rocket mail: e-mail use by U.S. humanities faculty." *Journal of Academic Librarianship*, 23(2), 109-117.
28. Brown, J. and Sime, J. (1981) "A methodology for accounts.", in: *Social method and social life*, edited by M. Brenner. London: Academic Press, pp. 159-188.
29. www.jegsworks.com/Lessons/lesson8/lesson8-1.htm
30. www.computerweekly.com/articles/clist.aspx
31. wiki.answers.com/Q/What_are_types_of_Computer_Software
32. en.wikipedia.org/wiki/Software

33. www.trybestsoft.com
34. Burton, P.F. (1994) "Electronic mail as an academic discussion forum." *Journal of Documentation*, 50(2), 99-110.
35. Case, D.O. (1986) "Collection and organization of written information by social scientists and humanists: A review and exploratory study." *Journal of Information Science*, 12, 97-104.
36. www.webopedia.com/TERM/O/operating_system.html
37. www.developers.net
38. www.computerhope.com/os.htm
39. www.dmoz.org/Computers/Software/Operating_Systems/Open_Source
40. www.techweb.com/encyclopedia/defineterm.jhtml
41. www.gnu.org
42. citeseer.ist.psu.edu/593632.html
43. www.sun.com/software/solaris
44. www.windowsitlibrary.com/Content/356/04/1.html
45. www.handhelds.org
46. www.webopedia.com/TERM/A/application.html
47. trace.wisc.edu/docs/software_guidelines/toc.htm
48. www.tek.com/products/accessories/application_software
49. www.ncsa.uiuc.edu/.../IBMp690/IBM/usr/share/man/info/en_US/a_doc_lib/aixbman/admnconc/vmm_overview.htm
50. www.softjamboree.com/category-Information-Management.html
51. www.cerlim.ac.uk/edner/dissem/c1.doc
52. www.ncsi.iisc.ernet.in/raja/opendl/free-software.htm

3
Program and Programming Languages

COMPUTER PROGRAM

In computing, a computer program is a specific set of ordered operations for a *computer* to perform. In the modern computer the program contains a one-at-a-time sequence of instructions that the computer follows. Typically, the program is put into a storage area accessible to the computer. The computer gets one instruction and performs it and then gets the next instruction.

Programs can be characterized of two types such as interactive or batch in terms of what drives them and how continuously they run. An interactive program receives data from an interactive user. A batch program runs and does its work, and then stops. Batch programs can be started by interactive users who request their interactive program to run the batch program. A command interpreter or a Web *browser* is an example of an interactive program. Print jobs are also batch programs.

Computer programs are written by using some kind of computer language. The language statements are the *source program*. The source program is compiled with a special program called a language compiler and the result is called an *object program*. The object program is also known as *object module* and *compiled program*. The object program contains the string of 0s and 1s in *machine language* with that the logic processor works with.

The machine language of the computer is constructed by the language compiler with an understanding of the computer's logic architecture, that includes the set of possible computer instructions and the length of an instruction.

Basically computer program is a file or collection of files that instruct a computer what to do. The *computer* performs functions based upon the instructions in the program. Computer programs are rarely used in a stand-alone fashion. A computer program is one type of software, other types of software include configuration files that store user preferences, linked or shared libraries, and user files which contain data.

Computer programs are either written in machine language, which is directly readable by the computer's CPU yet difficult for humans to understand, or in a high-level programming language which is similar to human languages, but must be converted to machine language in order to run. The task of converting the high-level languages to machine language is usually done by a compiler (another program), which is written in an intermediate-level language that was likely considered a high-level language at the time it was written. Thus as new, more powerful programming languages are developed they can be used to develop newer, yet more powerful languages. Usually, computer programs that are acquired by the end user are already converted into machine language and the code that created it cannot be examined or altered.

Infact, a program is a sequence of instructions that specifies how to perform a computation.

Examples are as following in the perspectives of computing manner :

* mathematical

- — solving a system of equations
- — finding the roots of a polynomial

* symbolic computation

- — searching and replacing text in a document
- — interpreting a program

Basic Instructions

Instructions (or commands, or statements) look different in different programming languages, but a few basic instructions appear in almost every language:

* input

— To get data from the keyboard, or a file, or some other device.

* output

— To display data on the screen or send data to a file or other device.

* math

— To perform basic mathematical operations like addition and multiplication.

* conditional execution

— To check for certain conditions and execute the appropriate sequence of statements.

* repetition

— To perform some action repeatedly, usually with some variation.

All programs consist of such instructions. Programming is *the process of breaking a large, complex task up into smaller and smaller subtasks until eventually the subtasks are simple enough to be performed with one of the simple instructions.*

What is debugging?

Programming errors are called **bugs**. The process of tracking the programming errors down and correcting them is called **debugging**.

There are three kinds of program errors:

Syntax errors

Syntax refers to the structure of the program and rules about that structure.

In English, a sentence must begin with a capital letter and end with a period.

* this sentence contains a syntax error.
* So does this one

A Python example is mismatched parenthesis:

```
(5/2))
^
```

SyntaxError: invalid syntax

Run-time errors

* Occur when the particular faulty line is executed.
* Called **exceptions** because something exceptional (and usually bad) has happened.
* Execution of the program is normally terminated at this point.
* Information about the current state of the program is printed.

Logic or semantic errors

In this case the program runs successfully, in the sense that it does not generate any error messages, but it doesn't do what someone wants it to do. Instead it is doing what someone told semantically it to do! Debugging logical errors requires to work backwards from the observed output (if any) to determine what the program is actually doing internally.

COMPUTER PROGRAMMING

Writing software (computer programs) is describing how to do something. In its simplest form, it is a lot like writing down the steps it takes to do something - a *process*.

An important reason to consider learning a bit about how to program is that the concepts underlying it will be valuable, regardless of whether or not one go on to make a career out of it. It does exactly what is told it to do, which is not necessarily what is wanted.

Algorithm

An algorithm is a procedure for solving a problem. The word derives from the name of the mathematician, Mohammed ibn-Musa al-Ḳhwarizmi, who was part of the royal court in Baghdad and who lived from about 780 to 850. A computer *program* can be viewed as an elaborate algorithm. In mathematics and computer science, an algorithm usually means a small procedure that solves a recurrent problem. To make a computer do anything, a *computer program* is to be written. To write a computer program, tell the computer, step by step, exactly what it has to do. The computer then "executes" the program, following each step mechanically, to accomplish the end goal. The algorithm is the basic technique used to get the job done..

OBJECT-ORIENTED PROGRAMMING

Object-oriented programming (OOP) is a programming language model organized around "objects" rather than "actions" and data rather than logic. Historically, a program has been viewed as a logical procedure that takes input data, processes it, and produces output data. The programming challenge was seen as how to write the logic, not how to define the data. Object-oriented programming takes the view that what we really care about are the objects we want to manipulate rather than the logic required to manipulate them.

The first step in OOP is to identify all the objects someone want to manipulate and how they relate to each other, an exercise often known as *data modeling*. Once someone've identified an *object*, generalize it as a *class* of objects and define the kind of data it contains and any logic sequences that can manipulate it. Each distinct logic sequence is known as a *method*. A real instance of a class is called an "object" or, in some environments, an "instance of a class." The object or class instance is what is run in the computer. Its methods provide computer instructions and the class object characteristics provide relevant data.

The concepts and rules used in object-oriented programming provide these important aspects :

* The concept of a data class makes it possible to define subclasses of data objects that share some or all of the main class characteristics. This property of OOP forces a more thorough data analysis, reduces development time, and ensures more accurate coding.
* Since a class defines only the data it needs to be concerned with the code will not be able to accidentally access other program data. This characteristic of *data hiding* provides greater system security and avoids unintended data corruption.
* The definition of a class is reuseable not only by the program for which it is initially created but also by other object-oriented programs.
* The concept of data classes allows a programmer to create any new *data type* that is not already defined in the language itself.

One of the first object-oriented computer languages was called *Smalltalk*. Today *C++* and *Java* are the most popular object-oriented languages. The Java programming language is designed for use in distributed applications on networks and the Internet.

FORMAL AND NATURAL LANGUAGES

Natural Languages

Natural Languages are which people speak, like English, Spanish and French. These are not designed by people; they evolved naturally.

Formal Languages are designed for specific applications:

1. mathematicians
   — mathematical notation is a formal language for

denoting relationships between numbers and symbols

2. chemists
   — chemical notation is a formal language to represent chemical structure of molecules
3. computer scientists
   — *programming languages are formal languages that have been designed to express computations*

Formal languages tend to have strict rules about syntax.

* 3+3=6 is a syntactically correct mathematical statement
* H_2O is a syntactically correct chemical name

Two types of syntax rules are found :

* Token rules
  — tokens are basic elements of the language
  — words
  — numbers
  — chemical elements
* Structure rules
  — the way tokens are arranged in a statement

The process of determining the structure of a sentence in a natural or formal language is called **parsing**.

* Example: "The other shoe fell"
  — "the other shoe" is the subject
  — "fell" is the verb

After parsing a sentence someone can determine the meaning or **semantics** of the sentence.

Formal and natural languages share:

* tokens
* structure
* syntax
* semantics

Formal and natural languages differ on:

* ambiguity
  — Natural languages are full of ambiguity.
  — Formal languages are designed to be unambiguous. Each statement has exactly one meaning.
* redundancy
  — Natural languages are redundant to make up for ambiguity and reduce miscommunications.
  — Formal languages are less redundant and more concise.
* literalness
  — Natural languages are full of idiom and metaphor:
  — Formal languages mean exactly what they say.

The difference between formal and natural languages has been compared to the difference between prose and poetry.

PROGRAMMING LANGUAGE

A **programming language** is an *artificial language* that can be used to *control* the behavior of a machine, particularly a *computer*. Programming languages, like *human languages*, are defined through the use of *syntactic* and *semantic* rules, to determine structure and meaning respectively.Programming languages are used to facilitate communication about the task of organizing. Sometimes the term "programming language" is restricted to those languages that can express *all* possible algorithms; sometimes the term "*computer language*" is used for more limited artificial languages.

Certain important aspects are as following :

Function : A programming language is a language used to write *computer programs*, which instruct a *computer* to perform some kind of *computation*.

Target : Programming languages allow humans to communicate instructions to machines. In some cases, programming languages are used by one program or machine to program another; *PostScript* source code, for example, is frequently generated programmatically to control a *computer printer* or display.

Constructs : Programming languages may contain constructs for defining and manipulating *data structures* or for controlling the *flow of execution.*

Purpose

The most important purpose of programming languages is to provide instructions to a computer. Programming languages require a greater degree of precision and completeness. When using a natural language to communicate with other people, human can be ambiguous and make small errors, and still expect their intent to be understood. Computers do exactly what they are told to do, and cannot understand the code the programmer "intended" to write.

Many languages have been designed from scratch, altered to meet new needs, combined with other languages, and eventually fallen into disuse. The need for different computer languages arises from the variety of contexts in which languages are used:

* Programs range from tiny scripts written by individual hobbyists to huge systems written by hundreds of programmers.
* Programmers range in expertise from novices who need simplicity above all else, to experts who may be comfortable with considerable complexity.
* Programs must balance speed, size, and simplicity on systems ranging from *microcontrollers* to *supercomputers.*
* Programs may be written once and not change for generations, or they may undergo nearly constant modification.
* Finally, programmers may simply differ in their tastes: they may be accustomed to discussing problems and expressing them in a particular language.

The common trend in the development of programming

languages has been to add more ability to solve problems using a higher level of *abstraction*. Features have been added to newly developed programing languages that let programmers express ideas that are removed from simple translation into underlying hardware instructions. Now programmers are less tied to the needs of the computer, their programs can do more computing with less effort from the programmer. This lets them write more programs in the same amount of time. The use of a formal language is essential to prevent the introduction of meaningless constructs.

1. First-generation *programming language*

A **first-generation** *programming language* is a machine-level programming language. It consists of 1s and 0s. **1GL** or first-generation language was (and still is) *machine language* or the level of instructions and data that the processor is actually given to work on (which in conventional computers is a string of 0s and 1s). First and second generation languages during the period of 1950-60, which to many experienced programmers will say are machine and assembly languages. Programming language history really began with the work of Charles Babbage in the early nineteenth century who developed automated calculation for mathematical functions. Further developments in early 1950 brought us machine language without interpreters and compilers to translate languages. Micro-code is an example of the first generation language residing in the CPU written for doing multiplication or division. Computers then were programmed in binary notation that was very prone to errors. A simple algorithm resulted in lengthy code. This was then improved to mnemonic codes to represent operations. Originally, no translator was used to compile or assemble the first-generation language. The first-generation programming instructions were entered through the front panel switches of the computer system.

The main benefit of programming in a first-generation programming language is that the code that a user writes can run very fast and efficiently since it is directly executed by the *CPU*, but machine language is somewhat more difficult to learn than higher generational programming languages, and it is far more difficult to edit if errors occur, or, for example, if instructions need to be added into memory at some location, then all the instructions after the insertion point need to be moved down to make room in memory to accommodate the new instructions. Doing so on a front panel with switches can be very difficult. Furthermore *portability* is significantly reduced - in order to transfer code to a different computer it needs to be completely rewritten since the machine language for one computer could be significantly different from another computer. Architectural considerations make portability difficult too. For example, the number of *registers* on one CPU architecture could differ from those of another.

1GL is mainly used on now very ancient computers, machine level programming still finds a use in several areas of modern programming. First and foremost, any native-code compiler creates machine language. This is done without user interaction, usually from a higher-level language as *Fortran*, *C/C++* or *Pascal*, often with intermediate *byte code* or *assembly* code. Another use is for so-called *virtual machines*. In essence, each virtual machine just creates a translation bridge between machine code and byte code. The byte code is the same across all platforms and the translator module of the virtual machine then translates *just-in-time* each byte to the corresponding native machine instruction.

Second-generation Programming Language

2GL or second-generation language is *assembler* (sometimes called "assembly") language. A typical 2GL instruction looks like this:

ADD 12, 8

An assembler converts the assembler language statements into machine language.

A second-generation programming language is a term

usually used to refer to some form of *assembly language*. Unlike first-generation programming languages, the code can be read and written fairly easily by a human, but it must be converted into a machine readable form in order to run on a computer. The conversion process is simply a mapping of the assembly language code into binary *machine code* (the *first-generation language*). The language is specific to and dependent on a particular processor family and environment. Since it is a one-to-one mapping to the native language of the target processor it has significant speed advantages, but it requires more programming effort and is difficult to use effectively for large applications. Symbolic addresses allowed programmers to represent memory locations, variables and instructions with names. Programmers now had the flexibility not to change the addresses for new locations of variables whenever they are modified. This kind of programming is still considered fast and to program in machine language required high knowledge of the CPU and machine's instruction set. This also meant high hardware dependency and lack of portability. Assembly or machine code could not run on different machines. The primary niche for these languages are in kernels, device drivers, and system libraries. Compilers usually use these as intermediate languages between a higher level language and machine code. Besides those, they are almost never used directly.

Third-Generation Programming Languages

3GL or third-generation language is a "high-level" programming language, such as *PL/I*, *C*, or *Java*. Java language statements look like this:

```
public boolean handleEvent (Event evt) {
switch (evt.id) {
case Event.ACTION_EVENT: {
if ("Try me" .equald(evt.arg)) {
```

A *compiler* converts the statements of a specific high-level programming language into machine language. (In the case of

Java, the output is called *bytecode*, which is converted into appropriate machine language by a Java *virtual machine* that runs as part of an operating system platform.) A 3GL language requires a considerable amount of programming knowledge. Throughout the early 1960's till 1980 saw the emergence of the third generation programming languages. Languages like ALGOL 58, 60 and 68, COBOL, FORTRAN IV, ADA and C are examples of this and were considered as high level languages. Most of the languages had compilers and the advantage of this was speed. Independence was another factor as these languages were machine independent and could run on different machines. The advantages of high level languages include the support for ideas of abstraction so that programmers can concentrate on finding the solution to the problem rapidly, rather than on low-level details of data representation. The comparative ease of use and learning, improved portability and simplified debugging, modifications and maintenance led to reliability and lower software costs. COBOL (COmmon Business-Oriented Language), a business data processing language is an example of a language constantly improving over the decades. It started out with a language called FLOWMATIC in 1955 and this language influenced the birth of COBOL-60 in 1959. Over the years, improvements were done to this language and COBOL 61, 65, 68, 70 were developed, being recognised as a standard in 1961. Now the new COBOL 97 has included new features like Object Oriented Programming to keep up with current languages. One good possible reason for this is the fact that existing code is important and to totally develop a new language from start would be a lengthy process. This also was the rationalisation behind the developments of C and C++.

These languages were mostly created following sequential procedural operations and code executed using branches and loops. Although the syntax between these languages were different but they shared similar constructs and were more readable by programmers and users compared to assembly languages. Some languages were improved over time and some were influenced by previous languages, taking the desired features thought to be good and discarding unwanted ones. New features were also added to the desired features to make the language more powerful.

The third generation languages are less processor dependent than lower level languages. An advantage in languages like C++ is that it gives the programmers a lot of control over how things are done in creating applications. This control however calls for more in depth knowledge of how the operating system and computer works. Many of the real programmers now still prefer to use these languages despite the fact the programmer having to devote a substantial professional effort to the leaning of a new complicated syntax which sometimes have little relation to human-language syntax even if it is in natural language.

Fourth-Generation Prog ramming Language

4GL or fourth-generation language is designed to be closer to natural language than a 3GL language. Languages for accessing databases are often described as 4GLs. A 4GL language statement might look like this:

EXTRACT ALL CUSTOMERS WHERE "PREVIOUS PURCHASES" TOTAL MORE THAN $1000

Features evident in fourth generation languages quite clearly are that it must be user friendly, portable and independent of operating systems, usable by non-programmers, having intelligent default options about what the user wants and allowing the user to obtain results fasts using minimum requirement code generated with bug-free code from high-level expressions, which was not possible using COBOL or PL/I. Standardisation however, in early stages of evolution can inhibit creativity in developing powerful languages for the future. Examples of this generation of languages are IBM's ADRS2, APL, CSP and AS, Power Builder, Access.

While third generation languages often followed procedural code, meaning the language performs functions defined in specific procedures on how something is done. In comparison, most fourth generation languages are nonprocedural. A disadvantage with fourth generation languages was they were slow compared to compiled languages and they also lacked control. Powerful languages of the future will combine procedural code and nonprocedural statements together with the flexibility of interactive screen applications, a powerful way of developing application. The need for this kind of languages is in line with minimum work and skill concept, point and click, programmers who are end users of software applications designed using third generation languages unseen by the commercial users. It is found that programmers whose primary interests are programming and computing use third generation languages and programmers who use the computers and programs to solve problems from other applications are the main users of the fourth generation languages.

Fifth-Generation Programming Language

5GL or fifth-generation language is programming that uses a visual or graphical development interface to create source language that is usually compiled with a 3GL or 4GL language compiler. Visual programming allows someone to easily envision *object-oriented programming class* hierarchies and drag icons to assemble program components. Fifth generation languages like PROLOG were developed in 90s referring to systems used in the field of *artificial intelligence*, *fuzzy logic* and *neural networks*. This means computers can in the future have the ability to think for themselves and draw their own inferences using programmed information in large databases. In fact, these databases programmed in a specialised area of study would show a significant expertise greater than humans. Improvements in the fourth generation languages carried features such as users did not need any programming knowledge. Little or no coding and computer aided design with graphics provides an easy to use product that can generate new applications.

VARIOUS COMPUTER LANGUAGES

BASIC

BASIC (standing for Beginner's All Purpose Symbolic Instruction Code) is a system developed in 1964. It was implemented for the G.E.225. It was meant to be a very simple language to learn and also one that would be easy to translate. Furthermore, the designers wished it to be a stepping-stone for students to learn on of the more powerful languages such as FORTRAN or ALGOL. Variable names of one character. No ELSE statement. The language was originally only available in interpreted versions, which were mostly running on mainframes and minis but with the advent of the personal computer in the mid-seventies many different versions began to appear, most holding to the Dartmouth spec, with extensions. Some didn't. A dialect running under CP/M on a North Star Horizon (1980) used ! as a statement seperator, and required to define strings as arrays (bit like 'C').

C

C has the strength that it is quite ubiquitous, and has simple enough requirements for the underlying "system model" that it is useful for implementing operating systems. C was developed in 1972 by Dennis Ritchie while working at Bell Labs in New Jersey. The transition in usage from the first major languages to the major languages of today occurred with the transition between Pascal and C. Its direct ancestors are B and BCPL, but its similarities to Pascal are quite obvious. All of the features of Pascal, including the new ones such as the CASE statement are available in C. C uses pointers extensively and was built to be fast and powerful at the expense of being hard to read. But because it fixed most of the mistakes Pascal had, it won over former-Pascal users quite rapidly.

C was developed for the new Unix system being created at the same time. Because of this, C and Unix go hand in hand.

Unix gives C such advanced features as dynamic variables, multitasking, interrupt handling, forking, and strong, low-level, input-output. Because of this, C is very commonly used to program operating systems such as Unix, Windows, the MacOS, and Linux. In the late 1970's and early 1980's, a new programing method was being developed. It was known as Object Oriented Programming, or OOP. Objects are pieces of data that can be packaged and manipulated by the programmer. Based on this method extensions to C were developed known as "C With Classes." This set of extensions developed into the full-featured language C++.

ALGOL

ALGOL was the first second-generation programming language and its characteristics are typical of the entire generation. First consider the data structures, which are very close to first generation structures. In ALGOL 60 the block structure was introduced: the ability to create blocks of statements for the scope of variables and the extent of influence of control statements. Along with that, two different means of passing parameters to subprograms; call by value and call by name. Structured control statements: if - then - else and the use of a general condition for iteration control was also features, as was the concept of recursion: the ability of a procedure to call itself. Several of ALGOL characteristics are listed below:

1. **Dynamic Arrays** : one for which the subscript range is specified by variables so that the size of the array is set at the time storage is allocated.
2. **Reserved Words** : the symbols used for keywords are not allowed to be used as identifiers by the programmer.
3. **User defined data types** : allow the user to design data abstractions that fit particular problems very closely.

ALGOL was used mostly by research computer scientists in the United States and in Europe. Its use in commercial applications was hindered by the absence of standard input/ output facilities in its description and the lack of interest in the language by large computer vendors. ALGOL 60 did however become the standard for the publication of algorithms and had a profound effect on future language development

Pascal

Pascal was begun in 1968 by Niklaus Wirth. Its development was mainly out of necessity for a good teaching tool. In the beginning, the language designers had no hopes for it to enjoy widespread adoption. Instead, they concentrated on developing good tools for teaching such as a debugger and editing system and support for common early microprocessor machines which were in use in teaching institutions. Pascal was designed in a very orderly approach, it combined many of the best features of the languages in use at the time, COBOL, FORTRAN, and ALGOL. While doing so, many of the irregularities and oddball statements of these languages were cleaned up, which helped it gain users (Bergin, 100-101). The combination of features, input/output *and* solid mathematical features, made it a highly successful language. Pascal also improved the "pointer" data type, a very powerful feature of any language that implements it. It also added a CASE statement that allowed instructions to branch like a tree in such a manner:

```
CASE expression OF
    possible-expression-value-1:
        statements to execute...
    possible-expression-value-2:
        statements to execute...
END
```

Pascal also helped the development of dynamic variables, which could be created while a program was being run, through the NEW and DISPOSE commands. However, Pascal did not implement dynamic arrays, or groups of variables, which proved to be needed and led to its downfall (Bergin, 101-102). Wirth later created a successor to Pascal, Modula-2, but by the time it appeared, C was gaining popularity and users at a rapid pace.

COBOL

COBOL is a high-level programming language first developed by the CODASYL Committee (**Co**nference on **Da**ta **Sy**stems **L**anguages) in 1960. Since then, responsibility for developing new COBOL standards has been assumed by the American National Standards Institute (ANSI). Three ANSI standards for COBOL have been produced: in 1968, 1974 and 1985. A new COBOL standard introducing object-oriented programming to COBOL, is due within the next few years. As the expanded acronym indicates, COBOL is designed for developing business, typically file-oriented, applications. It is not designed for writing systems programs. For instance somebody would not develop an operating system or a compiler using COBOL. COBOL is a simple language (no pointers, no user defined functions, no user defined types) with a limited scope of function. It encourages a simple straightforward programming style. Curiously enough though, despite its limitations, COBOL has proven itself to be well suited to its targeted problem domain (business computing). Most COBOL programs operate in a domain where the program complexity lies in the business rules that have to be encoded rather than in the sophistication of the data structures or algorithms required. And in cases where sophisticated algorithms are required COBOL usually meets the need with an appropriate verb such as the SORT and the SEARCH.

FORTRAN

The first FORTRAN compiler was a milestone in the history of computing, at that time computers had very small memories (on the order of 15KB, it was common then to count memory capacities in bits), they were slow and had very primitive operating systems (if they had them at all). At those days it seemed that the only practical way is to program in assembly language. FORTRAN II (1958) was a significant improvement, it added the capability for separate compilation of program modules, assembly language modules could also be 'linked loaded' with FORTRAN modules. FORTRAN III (1958) was never released to the public. It made possible using assembly language code right in the middle of the FORTRAN code.

Such "inlined" assembly code can be more efficient, but the advantages of an HLL are lost (e.g. portability, ease of use). FORTRAN IV (1961) was a 'clean up' of FORTRAN II, improving things like the implementation of the COMMON and EQUIVALENCE statements, and eliminating some machine-dependant language irregularities. A FORTRAN II to FORTRAN IV translator was used to retain backward compatibility with earlier FORTRAN programs.

LISP

Lisp offers the merits of representing both code and data as lists, providing higher order functions, and providing the ability to manipulate the representation of code at compile time via macros. As a programming language, LISP is characterized by the following ideas:

1. Computing with symbolic expressions rather than numbers.
2. Representation of symbolic expressions and other information by list structure in computer memory.
3. Representation of information on paper, from keyboards and in other external media mostly by multi-level lists and sometimes by S-expressions. It has been important that any kind of data can be represented by a single general type.
4. A small set of selector and constructor operations expressed as functions, i.e. car, cdr and cons.
5. Composition of functions as a tool for forming more complex functions.
6. The use of conditional expressions for getting branching into function definitions.
7. The recursive use of conditional expressions as a sufficient

tool for building computable functions.

8. The use of lambda-expressions for naming functions.
9. The storage of information on the property lists of atoms. The conditional expression interpretation of Boolean connectives.
10. The LISP function eval that serves both as a formal definition of the language and as an interpreter.
11. Garbage collection as the means of erasure.
12. Minimal requirements for declarations so that LISP statements can be executed in an on-line environment without preliminaries.
13. LISP statements as a command language in an on-line environment.

PERL

Perl allows building compact programs that quite powerfully manipulate text, and links to *huge* libraries of utilities. Perl has often been described as the "duct tape of the Internet," because it is most often used as the engine for a web interface or in scripts that modify configuration files. It has very strong text matching functions which make it ideal for these tasks. Perl was developed by Larry Wall in 1987 because the Unix sed and awk tools (used for text manipulation) were no longer strong enough to support his needs. Perl stands for Practical Extraction and Reporting Language or Pathologically Eclectic Rubbish Lister.

Python

Python brings together a somewhat "stricter" data model than Perl with a simpler syntax. It Combines ideas from ABC, C, Modula-3, and ICON. It bridges the gap between C and shell programming, making it suitable for rapid prototyping or as an extension of C. Rossum wanted to correct some of the ABC problems and keep the best features. At the time, he was working on the AMOEBA distributed OS group, and was looking for a scripting language with a syntax like ABC but with the access to the AMOEBA system calls, so he decided to create a language that was extensible; it is OO and supports packages, modules, classes, user-defined exceptions, a good C interface, dynamic loading of C modules and has no arbritrary restrictions

ML

ML (metalanguage) uses "theorem-proving" concepts to validate that the data manipulations expressed in a function validly correspond to the function's "signature," which is is helpful in verifying correctness of code. A strict higher-order functional language with statically-checked polymorphic types, garbage collection and a formal semantics. It began as the metalanguage for the Edinburgh LCF proof assistant. (LCF="Logic for Computable Functions"). LCF ML was implemented in Stanford LISP. (Meta-language) is a family of advanced languages with (usually) functional control structures, strict semantics, a strict polymorphic type system and parameterized modules. The family includes; Standard ML, Lazy ML, Caml, Caml Light, and various research languages.

Prolog

Prolog uses backtracking and (seeming) indeterminism so that someone *describe* answers rather than indicating how to calculate them. This is very useful when someone know what answers someone want, but either don't care or don't know the sequencing of how the result is to be found. Prolog is a logical and a declarative programming language. The name itself, Prolog, is short for PROgramming in LOGic. Prolog's heritage includes the research on theorem provers and other automated deduction systems developed in the 1960s and 1970s. The other major influence on the nature of this first Prolog was that it was designed to facilitate natural language processing. Prolog is the major example of a fourth generation programming language supporting the declarative programming paradigm. The Japanese Fifth-Generation Computer Project, announced in 1981, adopted Prolog as a development language, and thereby focused considerable attention on the language and its capabilities.

Java

Java is a language that was originally created to be used to program toasters, microwave ovens, and other embedded systems. It combines:

1. A C/C++-like language structure;
2. Object oriented features in the form of *classes* and *methods* (multiple inheritance is specifically omitted);
3. An assumption of the use of garbage collection to manage dynamically-allocated memory;
4. A Java bytecode machine. Java compilers generate "machine code" in this bytecode form; every Java platform must then have a bytecode interpreter to execute the code. This provides platform independence at the cost of reduced speed of execution; "Just In Time optimizers" should soon reduce the performance loss that results from this approach.
5. A security model for executing code obtained over networks that tries to protect users against the ill effects of "rogue code."

Java technology was created as a programming tool in a small, closed-door project.. But creating a new language wasn't even the point of "the Green Project."

The secret "Green Team," fully staffed at 13 people, was chartered by Sun to anticipate and plan for the "next wave" in computing. Their initial conclusion was that at least one significant trend would be the convergence of digitally controlled consumer devices and computers.

Javascript

A scripting language developed by Netscape to enable Web authors to design interactive sites. Although it shares many of the features and structures of the full JAVA language, it was developed separately. JAVASCRIPT can interact with HTML source code, enabling Web authors to spice their sites with dynamic content. JAVASCRIPT is endorsed by a number of software companies, and it is an Open Source product that anyone can use without purchasing a license. It is supported by recent browsers from Netscape and MS; although IE supports a subset, which MS calls JSCRIPT. "Within HTML, JAVASCRIPT reigns supreme and is the default script. It has C++ and JAVA like structures and can port some of the C/ C++ and JAVA code. JAVASCRIPT basically extends the HTML language, but it's scope seems confined to the browser

PHP

PHP is a widely-used general-purpose scripting language that is especially suited for Web development and can be embedded into HTML. PHP file may contain text, HTML tags and scripts. Scripts in a PHP file are executed on the server. this is different from a script written in other languages like Perl or C — instead of writing a program with lots of commands to output HTML, someone write an HTML script with some embedded code to do something (in this case, output some text). The PHP code is enclosed in special start and end tags that allow to jump into and out of "PHP mode".

* PHP is a server-side scripting language, like ASP
* PHP scripts are executed on the server
* PHP supports many databases (MySQL, Informix, Oracle, Sybase, Solid, PostgreSQL, Generic ODBC, etc.)
* PHP is an open source software (OSS)
* PHP is free to download and use

What distinguishes PHP from something like client-side JavaScript is that the code is executed on the server. If someone was to have a script similar to the above on the server, the client would receive the results of running that script, with no way of determining what the underlying code may be. Someone can even configure the web server to process all the HTML files with PHP, and then there's really no way that users can tell what up the sleeve. The best things in using PHP are that it is extremely simple for a newcomer, but offers many advanced features for a professional programmer. Don't be afraid reading the long list of PHP's features. Someone can jump in, in a short time, and start writing simple scripts in a few hours.

Although PHP's development is focused on server-side scripting, someone can do much more with it.

OPEN SOURCE

Open source is a programming philosophy whereby all aspects of a *program* are open for review. The term usually refers to the source code of the program, but can also refer to files created by the program, which would be readable in other applications. Many *Linux* distributions contain exclusively open source *software*. The opposite of open source software is proprietary software, whereby the source code is available only to the developer of the software and to those who successfully *hack* it.

The availability of the source code is seen as a security advantage to many, and as a security liability to others. Those who oppose open source software claim that *hackers* can easily identify weaknesses in software by examining the source code. Those in favor of open source software note that there are many more programmers that benefit from identifying and fixing potential weaknesses than those who benefit from exploiting them. Therefore, potential security holes are repaired before hackers can exploit them. In the real world, open source software such as Linux and *Firefox* have an exploit-to-patch time of under 12 hours on average, with an average exploit-to-patch time of 23 days for proprietary products such as *Windows* and *Internet Explorer*. In fact, the Linux vs. Windows and Firefox vs. Internet Explorer debates are seen as the two main battlefields of open source vs. proprietary software. The main object of these debates is security.

Another advantage of paid, proprietary software is the

support contract that usually accompanies it. Those who purchase Microsoft Windows, for instance, are provided with free installation help by telephone, and can purchase additional support contracts. Linux, on the other hand, is usually downloaded free of charge, and those who provide the download are under no obligation to provide installation support for it. However, most Linux distributions have large community-based support via mailing lists and online forums, offering help not only in the installation of the product, but also in other areas as well. Others vendors, such as Red Hat, offer paid versions of their software that include support contracts similar to Microsoft's extended support.

REFERENCES

1. In mathematical terms, this means the programming language is *Turing-complete MacLennan, Bruce J. (1987). Principles of Programming Languages. Oxford University Press. ISBN 0-19-511306-3.*
2. *As of May 2006 The Encyclopedia of Computer Languages* by *Murdoch University, Australia* lists 8512 computer languages.
3. *ACM* SIGPLAN (2003). *Bylaws of the Special Interest Group on Programming Languages of the Association for Computing Machinery.* Retrieved on *2006-06-19., The scope of SIGPLAN is the theory, design, implementation, description, and application of computer programming languages - languages that permit the specification of a variety of different computations, thereby providing the user with significant control (immediate or delayed) over the computer's operation.*
4. Dean, Tom (2002). *Programming Robots. Building Intelligent Robots.* Brown University Department of Computer Science.
5. Adler, S. (2001). *Extensible Stylesheet Language (XSL) Version 1.0. W3C Recommendation 15 October 2001.* Cambridge, MA: W3C. Available at http://www.w3.org/TR/xsl.
6. Ankolekar, A., et al., (2001), "DAML-S: semantic markup for web services.", in: *The First Semantic Web Working Symposium*, pp. 411-430, Heidelberg: Springer-Verlag Heidelberg.
7. Barros, F., Goncalves, P. & Santos, T. (1998). "Providing context to web searches: the use of ontologies to enhance search engine's accuracy." *In Journal of the Brazilian Computer Society.* 5(2), 45-55.
8. Bellwood, T., Clément, L., Ehnebuske, D. Hately, A., Hondo, M., Husband, Y.L., Januszewski, K., Lee, S., McKee, B., Munter, J. & von Riegen, C. (2002)., *UDDI Version 3.0 Specification Open Draft Dated 03*

July 2002. UDDI.Org Available at http://uddi.org/pubs/UDDI-V3.00-Open-Draft-20020703.htm

9. www.mitpress.mit.edu/sicp/full-text/book/book.html
10. *www.britannica.com/eb/article-9025053/computer-program*
11. *www.businesslink.gov.uk/bdotg/action/layer?topicId=1073935242*
12. Berners-lee, T. (1998), "*Semantic web road map.*" *IW3C Design Issues.* Cambridge, MA: W3C. Cambridge, MA: W3C. Available at: http://www.w3.org/DesignIssues/Semantic.html.
13. Birrell, A. & Nelson, B. (1984), "Implementing remote procedure calls." *ACM Transactions on Computer Systems*, 2(1), 39-59.
14. *Boag, S., et al., XQuery 1.0: An XML Query Language, W3C Working Draft 30 April 2002.* Cambridge, MA: W3C. Available at http://www.w3.org/TR/xquery.
15. Box, D., Ehnebuske, D., Kakivaya, G., Layman, A., Mendelsohn, N., Nielsen, H.F., Thatte, S., and Winer, D. (2000) *Simple Object Access Protocol (SOAP) 1.1 W3C Note 08 May 2000* Cambridge, MA: W3C Available at: http://www.w3.org/TR/SOAP/.
16. Bray, T., Paoli, J., & Sperberg-McQueen, C.(eds) (1998), *Extensible markup language (XML) 1.0. W3C Recommendation 10-February-1998.* Cambridge, MA: W3C. Available at: http://www.w3.org/TR/1998/REC-xml-19980210.
17. Digital Equipment Corporation. *Information Technology - Database Language SQL (Proposed revised text of DIS 9075). ISO/IEC 9075:1992, Database Language SQL.*
18. en.wikipedia.org/wiki/Computer_program
19. *www.howstuffworks.com/program.htm*
20. *www.webopedia.com/TERM/O/program.html*
21. findarticles.com/p/articles
22. birg.cs.wright.edu/cs271/textbook/app-1.pdf
23. *www.bfoit.org/Intro_to_Programming/Programming.html*
24. mathbits.com/mathbits/compsci/Introduction/intro.htm
25. *www.phon.ucl.ac.uk/courses/spsci/matlab*
26. *www.cs.brown.edu/courses/cs015*
27. *www.microsoft.com/learning/syllabi/2667Afinal.msp*
28. The Charity Development Group (December 1996). *The CHARITY Home Page*. Retrieved on *2006-06-29., Charity is a categorical programming language..., All Charity computations terminate.*
29. IBM in first publishing PL/I, for example, rather ambitiously titled its manual *The universal programming language PL/I* (IBM Library; 1966). The title reflected IBM's goals for unlimited subsetting capability: *PL/I is designed in such a way that one can isolate subsets from it satisfying the*

requirements of particular applications. (*Encyclopaedia of Mathematics » P » PL/I. SpringerLink. Ada* and *UNCOL* had similar early goals.

30. Frederick P. Brooks, Jr.: *The Mythical Man-Month,* Addison-Wesley, 1982, pp. 93-94
31. Dijkstra, Edsger W. *On the foolishness of "natural language programming."* EWD667.
32. Perlis, Alan, *Epigrams on Programming.* SIGPLAN Notices Vol. 17, No. 9, September 1982, pp. 7-13.

4
DOS

DOS stands for *disk operating system*. The term *DOS* is most often used as shorthand for *MS-DOS (Microsoft disk operating system)*. Originally developed by Microsoft for IBM, MS-DOS was the standard operating system for IBM-compatible personal computers. The initial versions of DOS were very simple. Subsequent versions have become increasingly sophisticated as they incorporated features of minicomputer operating systems. DOS is still a 16-bit operating system and does not support multiple users or multitasking. Microsoft Windows helped alleviate some problems, but still, it sat on top of DOS and relied on DOS for many services. Even Windows 95 sat on top of DOS. Newer operating systems, such as Windows NT do not rely on DOS to the same extent, although they can execute DOS-based programs. In fact Microsoft DOS (Disk Operating System) is a command line user interface. MS-DOS 1.0 was released in 1981 for IBM computers and the latest version of MS-DOS is MS-DOS 6.22, which was released in 1994. While MS-DOS is not commonly used by itself today, it can be accessed from every version of Microsoft Windows by clicking Start / Run and typing "command" or by typing "CMD" in Windows NT, Windows 2000 or Windows XP.

FEATURES OF DOS

Multitasking

Initially MS-DOS was not designed to be a multi-user or multitasking operating system, but many attempts were made to add these capabilities. Terminate and Stay Resident (TSR) system calls were originally designed for device drivers and extensible plugins that enhanced or added features. Add-on environments like TopView and especially DESQview attempted to provide multitasking, and achieved some success when later combined with the virtual 8086 mode and virtual memory features of the Intel 80386 and later processors.

User interface

MS-DOS employs a command line interface and a batch scripting facility via its command interpreter, command.com. MS-DOS was designed so users could easily substitute a different command line interpreter. Beginning with version 4.0, MS-DOS included a file manager program with a quasi-graphical user interface (the DOS Shell) that featured menus, split windows, and program shortcuts using character mode graphics.

Batch file : What are these

Batch files are files which allow MS-DOS and Microsoft Windows users to create a lists of commands and/or programs to run once the batch file has been executed. For example, a batch file could be used to run frequently run commands, deleting a series of files, moving files, etc. Much like a shortcut, batch files could be used to run one or more commands and/or programs through the command line. Example of a very well known batch file is the autoexec.bat, which is a simple boot file loaded each time the computer is loaded on MS-DOS and early Windows computers. This batch file contains all the necessary commands and programs used to run MS-DOS and Windows each time the computer booted.

CREATING A BATCH FILE

MS-DOS users

To create a basic batch file in MS-DOS, the following steps gives an example of how to create a basic batch file.

1. Open an MS-DOS command window or get to MS-DOS. Additional information about doing this can be found on document CHDOS.
2. At the MS-DOS prompt, type **edit test.bat** and press enter.
3. Then, there should now be a blue screen. Within the screen, type:

   pause

   dir c:\windows

   dir c:\windows\system
4. Once the above three lines have been typed in, click File and choose exit; when prompted to save, click "Yes". Users who do not have a mouse cursor can accomplish this same task by pressing ALT+F to access the file menu, then pressing "X" to exit, and pressing enter to save changes.
5. When back at the MS-DOS prompt, type test and press enter. This will execute the test.bat file and begin running the file. Because the first line is pause, user will first be prompted to press a key. Once user presses a key the batch file will run line-by-line; in this case, listing the files in the windows and windows\system directories.

When user wishes to add more lines to this batch file he would simply type "edit test.bat" to edit the file again.

Microsoft Windows and other users

A Windows user can still use the above MS-DOS steps if they wish to create a batch file. If, however, being comfortable using Microsoft Windows or the operating system, one can use any text editor, such as notepad or wordpad, to create the batch files, as long as the file extension ends with .bat. Following is the example:

1. Click Start
2. Click Run
3. Type "notepad" and press enter.
4. Once notepad is open, type the below lines in the file or copy and paste the below lines into notepad.

   ```
   @echo off
   echo Hello this is a test batch file
   pause
   dir c:\windows
   ```

5. Click File and click Save; browse to where one wants to save the file. For the file name, type "test.bat", and if the version of Windows has a "Save as type" option, choose "All files", otherwise it will save as a text file. Once all of this has been done click the Save button and exit notepad.

Now, to run the batch file, simply double-click or run the file like any other program. Once the batch file has completed running it will close the window automatically.

COMMAND LINE

A prompt is where there the user types in a command, as opposed to using the mouse to perform a command. For example, a root MS-DOS command line is generally C:\>. Unlike a GUI Operating System, a command line Operating System only uses a keyboard to navigate.

Since command line operating systems require commands to be used, these types of operating systems can be difficult to learn for new users. Command line Operating System can be a very valuable resource and should not be ignored.

Various Commands:

* APPEND
  (External)
  APPEND ;
  APPEND [d:]path[;][d:]path[...]
  APPEND [/X:on|off][/path:on|off] [/E]

Displays or sets the search path for data files. DOS will search the specified path(s) if the file is not found in the current path.

* ASSIGN

(External)

ASSIGN x=y [...] /sta

* Redirects disk drive requests to a different drive.

ATTRIB

(External)

ATTRIB [d:][path]filename [/S]

ATTRIB [+R|-R] [+A|-A] [+S|-S] [+H|-H] [d:][path]filename [/S]

It sets or displays the read-only, archive, system, and hidden attributes of a file or directory.

* BACKUP

(External)

BACKUP d:[path][filename] d:[/S][/M][/A][/F:(size)] [/P][/D:date] [/T:time] [/L:[path]filename]

It makes a backup copy of one or more files. (In DOS Version 6, this program is stored on the DOS supplemental disk.

* BREAK

(Internal)

BREAK =on|off

When used from the DOS prompt or in a batch file or in the CONFIG.SYS file to set (or display) whether or not DOS should check for a Ctrl + Break key combination.

* BUFFERS

(Internal)

BUFFERS=(number),(read-ahead number)

When used in the CONFIG.SYS file to set the number of disk buffers (number) that will be available for use during data input. Also used to set a value for the number of sectors to be read in advance (read-ahead) during data input operations.

* CALL

(Internal)

CALL [d:][path]batchfilename [options]

It calls another batch file and then returns to current batch file to continue.

* CHCP

(Internal)

CHCP (codepage)

It displays the current code page or changes the code page that DOS will use.

* CHDIR

(Internal)

CHDIR (CD) [d:]path

CHDIR (CD)[..]

It displays working directory and/or changes to a different directory.

* CHKDSK

(External)

CHKDSK [d:][path][filename] [/F][/V]

It checks a disk and provides a file and memory status report.

* CHOICE

(Internal)

CHOICE [/C[:]keys] [/N][/S][/T[:]c,nn] [text]

When used to provide a prompt so that a user can make a choice while a batch program is running.

* CLS (Clear Screen)

(Internal)

CLS

Clears (erases) the screen.

* COMMAND

(External)

COMMAND [d:][path] [device] [/P][/E:(size)] [/MSG][/Y [/C (command) | /K (command)]

It starts a new version of the DOS command processor (the program that loads the DOS Internal programs).

* COMP

(External)

COMP [d:][path][filename] [d:][path][filename] [/A][/C][/D][/L][/N:(number)]

It compares two groups of files to find information that does not match. (See FC command).

* COPY

(Internal)

COPY [/Y | -Y] [/A][/B] [d:][path]filename [/A][/B] [d:][path][filename] [/V]

or

COPY [/Y | -Y][/A][/B] [d:][path]filename+[d:][path]filename[...][d:][path][filename] [/V]

Copies and appends files.

* COUNTRY

(Internal)

COUNTRY=country code,[code page][,][d:][filename]

When used in the CONFIG.SYS file to tell DOS to use country-specific text conventions during processing.

* CTTY

(Internal)

CTTY (device)

It changes the standard I/O (Input/Output) device to an auxiliary device.

* DATE

(Internal)

DATE mm-dd-yy

It displays and/or sets the system date.

* DBLSPACE

(External)

DBLSPACE / automount=drives

DBLSPACE /chkdsk [/F] [d:]

DBLSPACE /compress d: [/newdrive=host:] [/reserve=size] [/F]

DBLSPACE /create d: [/newdrive=host:] [/reserve=size] [/size=size]

DBLSPACE /defragment [d:]]/F]

DBLSPACE /delete d:

DBLSPACE /doubleguard=0|1

DBLSPACE /format d:

DBLSPACE [/info] [d:]

DBLSPACE /list

DBLSPACE /mount[=nnn] host: [/newdrive=d:]

DBLSPACE /ratio[=ratio] [d:] [/all]

DBLSPACE /size[=size] [/reserve=size] d:

DBLSPACE /uncompress d:

DBLSPACE /unmount [d:]

This program is available with DOS 6.0 that allows user to compress information on a disk.

* DEBUG

(External)

DEBUG [pathname] [parameters]

This MS-DOS utility is used to test and edit programs.

* DEFRAG

(External)

DEFRAG [d:] [/F][/S[:]order] [/B][/skiphigh [/LCD|/BW|/GO] [/H]

DEFRAG [d:] [/V][/B][/skiphigh] [/LCD]|/BW|/GO] [/H]

It optimizes disk performance by reorganizing the files on the disk.

* DEL (ERASE)

(Internal)

DEL (ERASE) [d:][path]filename [/P]

It deletes (erases) files from disk.

* DELOLDOS

(External)

DELOLDOS [/B]

It deletes all files from previous versions of DOS after a 5.0 or 6.0 installation.

* DELTREE

(External)

DELTREE [/Y] [d:]path [d:]path[...]

It deletes (erases) a directory including all files and subdirectories that are in it.

* DEVICE

(Internal)

DEVICE=(driver name)

It is used in the CONFIG.SYS file to tell DOS which device driver to load.

* DEVICEHIGH

(Internal)

DEVICEHIGH=(driver name)

Like DEVICE, DEVICEHIGH is used in the CONFIG.SYS file to tell DOS which device driver software to use for devices; however, this option is used to install the device driver into the upper memory area.

* DIR

(Internal)

DIR [d:][path][filename] [/A:(attributes)] [/O:(order)] [/B][/C][/CH][/L][/S][/P][/W]

It displays directory of files and directories stored on disk.

* DISKCOMP

(External)

DISKCOMP [d:] [d:][/1][/8]

It compares the contents of two diskettes.

* DISKCOPY

(External)

DISKCOPY [d:] [d:][/1][/V][/M]

It makes an exact copy of a diskette.

* DOS

(Internal)

DOS=[high | low],[umb | noumb]

It is used in the CONFIG.SYS file to specify the memory location for DOS. It is used to load DOS into the upper memory area and to specify whether or not the upper memory blocks will be used.

* DOSKEY

(External)

DOSKEY [reinstall] [/bufsize=size][/macros][/history][/insert | /overstrike] [macroname=[text]]

It loads the Doskey program into memory which can be used to recall DOS commands so that user can edit them.

* DOSSHELL

(External)

DOSSHELL [/B] [/G:[resolution][n]] | [/T:[resolution][n]]

It initiates the graphic shell program using the specified screen resolution.

* DRIVPARM

(Internal)

DRIVPARM= /D:(number) [/C] [/F:(form factor)] [/H:(number)] [/I][/N][/S:(number)] [/T:(tracks)]

It is used in the CONFIG.SYS file to set parameters for a disk drive.

* ECHO

(Internal)

ECHO on | off

ECHO (message)

It displays messages or turns on or off the display of commands in a batch file.

* EDIT

(External)

EDIT [d:][path]filename [/B][/G][/H][/NOHI]

It starts the MS-DOS editor, a text editor used to create and edit ASCII text files.

* EMM386

(External)

EMM386 [on|off|auto] [w=on|off]

It enables or disables EMM386 expanded-memory support on a computer with an 80386 or higher processor.

* EXE2BIN

(External)

EXE2BIN [d:][path]filename [d:][path]filename

It converts .EXE (executable) files to binary format.

* EXIT

(Internal)

EXIT

It exits a secondary command processor.

* EXPAND

(External)

EXPAND [d:][path]filename [[d:][path]filename[. . .]]

It expands a compressed file.

* FASTHELP

(External)

FASTHELP [command][command] /?

It displays a list of DOS commands with a brief explanation of each.

* FASTOPEN

(External)

FASTOPEN d:[=n][/X]

It keeps track of the locations of files for fast access.

* FC

(External)

FC [/A][/C][/L][/Lb n][/N][/T][/W][number] [d:][path]filename [d:][path]filename

or (for binary comparisons)

FC [/B][/number] [d:][path]filename [d:][path]filename

It displays the differences between two files or sets of files.

* FCBS

(Internal)

FCBS=(number)

It is used in the CONFIG.SYS file to specify the number of file-control blocks for file sharing.

* FDISK

(External)

FDISK [/status]

It prepares a fixed disk to accept DOS files for storage.

* FILES

(Internal)

FILES=(number)

It is used in the CONFIG.Sys file to specify the maximum number of files that can be open at the same time.

* FIND

(External)

FIND [/V][/C][/I][/N] ÒstringÓ [d:][path]filename[...]

It finds and reports the location of a specific string of text characters in one or more files.

* FOR

(Internal)

FOR %%(variable) IN (set) DO (command)

or (for interactive processing)

FOR %(variable) IN (set) DO (command)

It performs repeated execution of commands (for both batch processing and interactive processing).

* FORMAT

(External)

FORMAT d:[/1][/4][/8][/F:(size)] [/N:(sectors)] [/T:(tracks)][/B|/S][/C][/V:(label)] [/Q][/U][/V]

It formats a disk to accept DOS files.

* GOTO

(Internal)

GOTO (label)

It causes unconditional branch to the specified label.

* GRAFTABL

(External)

GRAFTABL [(code page)]

GRAFTABL [status]

It loads a table of character data into memory (for use with a color/graphics adapter).

* GRAPHICS

(External)

GRAPHICS [printer type][profile] [/B][/R][/LCD][/PB:(id)] [/C][/F][/P(port)]

It provides a way to print contents of a graphics screen display.

* HELP

(External)

HELP [command] [/B][/G][/H][/NOHI]

It displays information about a DOS command.

* IF

(Internal)

IF [NOT] EXIST filename (command) [parameters]

IF [NOT] (string1)==(string2) (command) [parameters]

IF [NOT] ERRORLEVEL (number) (command) [parameters]

It allows for conditional operations in batch processing.

* INCLUDE

(Internal)

INCLUDE= blockname

It is used in the CONFIG.SYS file to allow user to use the commands from one CONFIG.SYS block within another.

* INSTALL

(Internal)

INSTALL=[d:][\path]filename [parameters]

It is used in the CONFIG.SYS file to load memory-resident programs into conventional memory.

* INTERLINK

(External)

INTERLINK [client[:]=[server][:]]

It connects two computers via parallel or serial ports so that the computers can share disks and printer ports.

* INTERSVR

(External)

INTERSVR [d:][...][/X=d:][...] [/LPT:[n|address]] [/COM:[n|address]][/baud:rate] [/B][/V]

INTERSVR /RCOPY

It starts the Interlink server.

* JOIN

(External)

JOIN d: [d:path]

JOIN d: [/D]

It allows access to the directory structure and files of a drive through a directory on a different drive.

* KEYB

(External)

KEYB [xx][,][yyy][,][d:][path]filename [/E][/ID:(number)]

It loads a program that replaces the support program for

U. S. keyboards.

* LABEL

(External)

LABEL [d:][volume label]

It creates or changes or deletes a volume label for a disk.

* LASTDRIVE

(Internal)

LASTDRIVE=(drive letter)

It is used in the CONFIG.SYS file to set the maximum number of drives that can be accessed.

* LOADFIX

(Internal)

LOADFIX [d:][path]filename [parameters]

It ensures that a program is loaded above the first 64K of conventional memory, and runs the program.

* LOADHIGH

(Internal)

LOADHIGH (LH) [d:][path]filename [parameters]

It loads memory resident application into reserved area of memory (between 640K-1M).

* MEM

(External)

MEM [/program|/debug|/classify|/free|/module(name)] [/page]

It displays amount of installed and available memory, including extended, expanded, and upper memory.

* MEMMAKER

(External)

MEMMAKER [/B][/batch][/session][/swap:d] [/T][/undo][/W:size1,size2]

It starts the MemMaker program, a program that lets user optimize computer's memory.

* MENUCOLOR

(Internal)

MENUCOLOR=textcolor,[background]

It is used in the CONFIG.SYS file to set the colors that will be used by DOS to display text on the screen.

* MENUDEFAULT

(Internal)

MENUDEFAULT=blockname, [timeout]

It is used in the CONFIG.SYS file to set the startup configuration that will be used by DOS if no key is pressed within the specified timeout period.

* MENUITEM

(Internal)

MENUITEM=blockname, [menutext]

It is used in the CONFIG.SYS file to create a start-up menu from which user can select a group of CONFIG.SYS commands to be processed upon reboot.

* MIRROR

(External)

MIRROR [d:]path [d:] path [...]

MIRROR [d1:][d2:][...] [/T(drive)(files)] [/partn][/U][/1]

It saves disk storage information that can be used to recover accidentally erased files.

* MKDIR

(MD) (Internal)

MKDIR (MD) [d:]path

It creates a new subdirectory.

* MODE

(External)

MODE n

MODE LPT#[:][n][,][m][,][P][retry]

MODE [n],m[,T]

MODE (displaytype,linetotal)

MODE COMn[:]baud[,][parity][,][databits][,][stopbits][,][retry]

MODE LPT#[:]=COMn [retry]

MODE CON[RATE=(number)][DELAY=(number)]

MODE (device) CODEPAGE PREPARE=(codepage) [d:][path]filename

MODE (device) CODEPAGE PREPARE=(codepage list) [d:][path]filename

MODE (device) CODEPAGE SELECT=(codepage)

MODE (device) CODEPAGE [/STATUS]

MODE (device) CODEPAGE REFRESH

It sets mode of operation for devices or communications.

* MORE

(External)

MORE < (filename or command)

(name)|MORE

It sends output to console, one screen at a time.

* MOVE

(Internal)

MOVE [/Y|/-Y] [d:][path]filename[,[d:][path]filename [...]] destination

It is moves one or more files to the location user specify. Can also be used to rename directories.

* MSAV

(External)

MSAV [d:] [/S|/C][/R][/A][/L][/N][/P][/F][/video][/mouse]

MSAV /video

It scans computer for known viruses.

* MSBACKUP

(External)

MSBACKUP [setupfile] [/BW|/LCD|/MDA]

It is used to backup or restore one or more files from one

disk to another.

* MSCDEX

(External)

MSCDEX /D:driver [/D:driver2. . .] [/E][/K][/S][/V][/L:letter] [/M:number]

It is used to gain access to CD-ROM drives (new with DOS Version 6).

* MSD

(External)

MSD [/B][/I]

MSD [/I] [/F[d:][path]filename [/P[d:][path]filename [/S[d:][path]filename

It provides detailed technical information about computer.

* NLSFUNC

(External)

NLSFUNC [d:][path]filename

It is used to load a file with country-specific information.

* NUMLOCK

(Internal)

NUMLOCK=on | off

It is used in the CONFIG.SYS file to specify the state of the NumLock key.

* PATH

(Internal)

PATH;

PATH [d:]path[;][d:]path[...]

It sets or displays directories that will be searched for programs not in the current directory.

* PAUSE

(Internal)

PAUSE [comment]

It suspends execution of a batch file until a key is pressed.

* POWER
  (External)
  POWER [adv:max | reg | min] | std | off]

  It is used to turn power management on and off, report the status of power management, and set levels of power conservation.

* PRINT
  (External)
  PRINT [/B:(buffersize)] [/D:(device)] [/M:(maxtick)] [/Q:(value] [/S:(timeslice)][/U:(busytick)] [/C][/P][/T] [d:][path][filename] [...]

  It queues and prints data files.

* PROMPT
  (Internal)
  PROMPT [prompt text] [options]

  It changes the DOS command prompt.

* RECOVER
  (External)
  RECOVER [d:][path]filename
  RECOVER d:

  It resolves sector problems on a file or a disk. (Beginning with DOS Version 6, RECOVER is no longer available).

* REM
  (Internal)
  REM [comment]

  It is used in batch files and in the CONFIG.SYS file to insert remarks (that will not be acted on).

* RENAME (REN)
  (Internal)
  RENAME (REN) [d:][path]filename [d:][path]filename

  It changes the filename under which a file is stored.

* REPLACE
  (External)

REPLACE [d:][path]filename [d:][path] [/A][/P][/R][/S][/U][/W]

It replaces stored files with files of the same name from a different storage location.

* RESTORE

(External)

RESTORE d: [d:][path]filename [/P][/S][/B:mm-dd-yy] [/A:mm-dd-yy][/E:hh:mm:ss] [/L:hh:mm:ss] [/M][/N][/D]

It restores to standard disk storage format files previously stored using the BACKUP command.

* RMDIR (RD)

(Internal)

RMDIR (RD) [d:]path

Removes a subdirectory.

* SCANDISK

(External)

SCANDISK [d: [d: . . .]|/all][/checkonly|/autofix[/nosave]|/custom][/surface][/mono][/nosummay]

SCANDISK volume-name[/checkonly|/autofix[/nosave]|/custom][/mono][/nosummary]

SCANDISK /fragment [d:][path]filename

SCANDISK /undo [undo-d:][/mono]

It starts the Microsoft ScanDisk program which is a disk analysis and repair tool used to check a drive for errors and correct any problems that it finds.

* SELECT

(External)

SELECT [d:] [d:][path] [country code][keyboard code]

It formats a disk and installs country-specific information and keyboard codes (starting with DOS Version 6, this command is no longer available).

* SET

(Internal)

SET (string1)=(string2)

It inserts strings into the command environment. The set values can be used later by programs.

* SETVER

(External)

SETVER [d:]:path][filename (number)][/delete][/quiet]

It displays the version table and sets the version of DOS that is reported to programs.

* SHARE

(External)

SHARE [/F:space] [/L:locks]

It installs support for file sharing and file locking.

* SHELL

(Internal)

SHELL=[d:][path]filename [parameters]

It is used in the CONFIG.SYS file to specify the command interpreter that DOS should use.

* SHIFT

(Internal)

SHIFT

It increases number of replaceable parameters to more than the standard ten for use in batch files.

* SORT

(External)

SORT [/R][/+n] < (filename)

SORT [/R][/+n] > (filename2)

It sorts input and sends it to the screen or to a file.

* STACKS

(Internal)

STACKS=(number),(size)

It is used in the CONFIG.SYS file to set the number of stack frames and the size of each stack frame.

* SUBMENU

(Internal)

SUBMENU=blockname, [menutext]

It is used in the CONFIG.SYS file to create a multilevel menu from which user can select start-up options.

* SUBST

(External)

SUBST d: d:path

SUBST d: /D

It substitutes a virtual drive letter for a path designation.

* SWITCHES

(Internal)

SWITCHES= [/K][/F][/N][/W]

It is used in the CONFIG.SYS file to configure DOS in a special way; for example, to tell DOS to emulate different hardware configurations.

* SYS

(External)

SYS [source] d:

It transfers the operating system files to another disk.

* TIME

(Internal)

TIME hh:mm[:ss][.cc][A|P]

It displays current time setting of system clock and provides a way for user to reset the time.

* TREE

(External)

TREE [d:][path] [/A][/F]

It displays directory paths and (optionally) files in each subdirectory.

* TYPE

(Internal)

TYPE [d:][path]filename

It displays the contents of a file.

* UNDELETE
(External)
UNDELETE [d:][path][filename] [/DT|/DS|/DOS]
UNDELETE [/list|/all|/purge[d:]|/status|/load|/U|/S[d:]|/Td:[-entries]]
It restores files deleted with the DELETE command.

* UNFORMAT
(External)
UNFORMAT d: [/J][/L][/test][/partn][/P][/U]
It is used to undo the effects of formatting a disk.

* VER
(Internal)
VER
It displays the DOS version number.

* VERIFY
(Internal)
VERIFY on|off
It turns on the verify mode; the program checks all copying operations to assure that files are copied correctly.

* VOL
(Internal)
VOL [d:]
It displays a disk's volume label.

* VSAFE
(External)
VSAFE [/option[+|-]...] [/NE][/NX][Ax|/Cx] [/N][/D][/U]
VSAFE is a memory-resident program that continuously monitors computer for viruses and displays a warning when it finds one.

* XCOPY
(External)
XCOPY [d:][path]filename [d:][path][filename] [/A][/

D:(date)] [/E][/M][/P][/S][/V][/W][Y\-Y]

It copies directories, subdirectories, and files.

MS-DOS ERRORS

Various MS-DOS errors are as follows :

Abort : Usually, user will get this error message if user is trying to read a disk that is not readable or if no disk is present.

Access denied : Usually meaning the file has an attribute on it. TIP: if need to delete the file, just deltree *FILE*

Bad command or file name : Usually, this is when user miss type a command or if DOS does not know the command.

Drive not ready error : Disk in the drive is not readable, such as a disk not being in the disk drive.

Duplicate file name or file not found : When attempting to rename a file, the file does not exist or there is already a file with that name.

File cannot be copied onto itself : An attempt to copy a file could not be completed because the file already exists or same source and destination.

File creation error : Directory could be full because of the DOS limitation of files in one directory; the file could also already exist or the file was not copied correctly

File not found : It means that, in the directory that one searched, it does not have that certain file or there are no files in the directory.

Insufficient disk space : Disk is full or, if one is trying to copy a file to a disk, the file is too big to fit on the disk.

Invalid directory : Directory is invalid or does not exist in the directory one is currently in.

Invalid drive specification : Drive does not exist on the computer or cannot be found.

Invalid file name or file not found : DOS cannot find the file specified in the current directory or the file has an invalid character in it.

Invalid media, track 0 bad or unusable : Format command cannot format the specified disk or the disk capacity is invalid or bad disk.

Invalid parameter : Typed a command correctly but specified the incorrect parameter, such as format /z.

Invalid Switch : The wrong slash was typed, such as cd/ when it is supposed to be cd\.

Non-System disk or disk error : When booting up the computer, this error will usually occur when there is a floppy disk that is not bootable in the computer. If there is no disk inside the computer, this could be an indication of an unrecognized hard drive, unformatted hard drive, or hard drive with missing system files.

Write protect : Tried to write to a disk that is write protected.

REFERENCES

1. www.answers.com/topic/dos
2. www.colorado.edu/geography/gcraft/tips/doshelp.html
3. www.fdisk.com/doslynx
4. w3.aces.uiuc.edu/AIM/CCSOcourses/DOS-rev.html
5. www.mdgx.com/secrets.htm
6. www.columbia.edu/~em36/wpdos
7. www.chebucto.ns.ca/~ak621/DOS/BatBasic.html
8. www.glue.umd.edu/~nsw/ench250/dostutor.htm
9. home.att.net/~gobruen/progs/dos_batch
10. www.computerhope.com/msdos.htm
11. www.webopedia.com/TERM/D/DOS.html
12. www.freedos.org
13. searchsecurity.techtarget.com/sDefinition
14. www.nos.org/htm/os1.htm
15. www.easydos.com/dosindex.html
16. searchsecurity.techtarget.com/sDefinition/0,,sid14_gci 213911,00.html
17. http://www.cknow.com/tutorcom/dos02_dosparts.htm
18. www.members.fortunecity.com/pcmuseum/dos.htm
19. www.inventors.about.com/library/weekly/aa033099. htm
20. www.lookwayup.com/lwu.exe/lwu/d?s=f&w= Microsoft%20disk%20operating%20system

5
Library Automation

Even though the question of library automation seems to be very fundamental it is essential to emphasize the library automation. Secondly, while justifying need for library automation more than cost-effectiveness the benefits derived by the library users become the major consideration. Since library does not happen to be an economic entity such benefits need to be looked at in a different perspective. To appreciate the advantages it becomes necessary to highlight the different levels of library automation. For convenience it can be visualized at four levels:

i. Library cataloging system
ii. House keeping operations and networking
iii. Development of CD-ROM library / products
iv. E-mail system and internet

The library catalogue or index to the collection forms the base for most of the library activities such as acquisition, reference, bibliographic service, inter-library loan etc. The users of library card catalogue will appreciate how fast the retrieval is, search and printing in automated environment. If the same system is available in network environment, users can have simultaneous access to the same database. From the library staff point of view the cumbersome job of printing the cards and their subsequent filing gets eliminated. Also, it conserves space and saves stationary.

The second level automation will be to use software which can handle all the house keeping operations of the library such as acquisition, circulation and serial control thus creating a network within the library or becoming part of the existing network of the institution. Networking of computers within an organization helps the users to browse the cataloguing system from any of the workstation/ terminal.

A very handy technology available for library is the CD-ROM products which can be considered at the third level. The development of CD-ROM collection not only conserves space but also provides multi-user access in network environment. There are many self-tutorial CD-ROMS available with multi-media effect. Libraries facing high incidence of mutilation of materials will benefit from such electronic products. Also people doing empirical research can download data and directly take it to other software platform for analysis and making graphical presentation. Other technology which libraries can make use of is the e-mail system. This not only reduces the recurring expenditure but also be effective and fast. Sending reminders for non-receipt of journals by e-mail has proved to be very cost-effective. In addition to this, sharing of resources among libraries become easy. Few public domain e-mail software are available and there will be no additional expenditure incurred. Another technology which has revolutionized the information world is the development of internet. Subscribers of internet, in addition to getting access to various public domain databases and services, will also get free e-mail and fax facility. Some publishers have started giving content pages of journals and libraries having subscription to such journals can also have full text of the articles. Many academic and research institutes have given free access to their working papers.

Library automation can be defined simply as the use of computer and networking technologies in the library.

AREAS OF LIBRARY AUTOMATION

Library automation can take place in any or all of the following areas :

* Automation of library functions
* Use of electronic resources within the library (e.g. CD-ROMs)
* Accessing remote electronic resources (e.g. the Internet)
* Office automation (e.g. word-processing, spreadsheets, databases, etc.)
* Patron services (e.g. computer laboratory, multimedia center)

OBJECTIVES OF LIBRARY AUTOMATION

Library automation can influence in various manners. Some of the areas are highlighted here :

* To improve the level of information service and quality of information products.
* To fulfill such needs that cannot be achieved by manual system:
    — Sharing of resources
    — Information that appears only in electronic format.

There are some aspects to be kept in mind regarding library automation:

* How the automation is to be fully implemented?
* What if circumstances changed so that a new system would be able to be fully used?
* Are there any plans to build a union catalog? Work as a consortium?
* What is collection size?
* What non-book materials does collection contain?
* Which modules are the library is most interested in?
* How would the library like to see automated functions used?

* How many terminals will be required initially? after 2-3 years?
* Would it make sense to put items on the software database with the aid of a bibliographic utility and then convert systems?
* What kind of systems support is available to the library?
* What classification system is to be used?
* What subject headings list is to be used?
* What bibliographic standards are to be followed/adapted?
* If any form of MARC records is to be used, which?
* Are the authorities interested in academic library materials only or stuff for public, corporate, government and school libraries as well?

STAGES IN LIBRARY AUTOMATION

Library automation is an multi-step and multi-task activity. So, it requires thorough planning. Though planning is time-consuming, but it is usually cost-effective because time spent planning reduces the amount of time required for system implementation. The various stages involved in library automation are as following:

Step 1: Understanding of existing library services and technology

* Identifying existing services and functions provided by the library
* Identifying existing technology being used in the library
* Collecting and organizing basic statistical data

Step 2: Assessing needs and setting priorities for automation.

* Who should be involved in planning?
* Identifying approaches to satisfy the needs
* Setting priorities
* Developing the preliminary budget

Step 3: Translating needs and priorities into specifications

* Designing specifications
* Preparing the Request for Proposal (RFP)

Step 4: Evaluating proposals and selecting a software

* Seeing system demonstrations
* Analyzing vendor responses
* Costs
* Final decision

Step 5: Implementing the system

* Contract negotiations
* Hardware and software installation
* Training

Step 6: Retrospective conversion and barcoding

BIBLIOGRAPHIC DATA CONVERSION

Bibliographic data conversion is done in following methods :

* Direct keying from shelflist
* Matching and extraction of records from a bibliographic database, for examples:
* OCLC (http://www.oclc.org/)
* from a school library catalog, e.g. TWGH Chen Zao Men College (http://www.twghczm.edu.hk/slsbin/webopac.exe)
* Converting from non-MARC format to MARC

BARCODING THE COLLECTION

Barcoding is the process by which a barcode label is attached to an item in the library's collection. Barcodes serve as a computerized accession number - a unique identifier that links a specific book, journal issue, compact disc, etc., to the item record that describes it.

Methods of barcoding the library collection are as following:

* **Smart barcodes :** Barcode number is pre-assigned to a

particular item record. The library software prints this barcode together with call number and/title of the item. Library staff finds that specific item on shelf and put the barcode label on it.

* **Dump barcodes :** Barcode number on the barcode label is not related to any item until that number is manually entered into the item record by the library staff. This is usually done by scanning in the number using a barcode reader.

USER DATA CONVERSION AND BARCODING

User Data Conversion and Barcoding is also an important task for a library. It can be done in the following steps :

* Converting users data
* Deciding the barcode numbering system
* Barcoding users ID card

Training

Training is associated with continuous implementation of automated services and functioning of library. It has various aspects associated with itself such as :

* Involving staff members at required levels in planning activities.
* Making effective use of training.
* Identifing required separate training programs if any and deciding who will be trained, and individuals who will be responsible for any ongoing training in the area.
* Using training tips to make in-house training effective.
* Developing a plan for training users.

MINIMUM INFRASTRUCTURE

One File Server

* to hold the library system and database
* for access by workstations via the Intranet or the Internet

One to Two Public Workstations

* for students/teachers to access the Online Public Access Catalog (OPAC), CD-ROM network, and/or the Internet

One Circulation Workstation

* for library staff to check-in and check-out items, etc.

One Staff Workstation

* for cataloging, acquisitions, reporting, etc.

TYPICAL CONFIGURATION FOR LIBRARY AUTOMATION

Based on requirements, library can opt for any combination of information technology configuration. The typical configuration of the system required for library automation is as following:

Server

* Pentium PC
* Sufficient RAM for serving the network
* Large and fast hard disk
* Network card
* Operating system
* Backup Device
* Optional: networked CD-ROM drives

Workstation

* Pentium PC
* Sufficient RAM
* Sufficient hard disk
* Network card
* Operating system
* Optional: CD-ROM drive and sound card

LIBRARY AUTOMATION SYSTEM SELECTION GUIDELINES

There are many kind of library automation systems

available in the market. Each one has its own speciality. It is a big challenge for any library how to select a system. The library automation system selection guidelines are as following:

1. The library automation software are developed and designed based on the best practices that are internationally adopted in the library profession. These include:
   * Adoption of MARC-based bibliographic record. Record can be imported, created, updated and exported using the MARC 21 and ISO 2709 standards.
   * Bibliographic and item information must be stored separately in two different types of record so that more than one item records can be attached to one bibliographic record.
   * Support of internationally adopted library standards, including ISBD, AACR2, subject heading scheme, classification scheme, etc.
   * Automation of library operations and activities, including circulation, public catalog searching, cataloging, ordering, serials control, and reporting.
2. The library automation software must be supported by a team that possesses library experience and qualification to ensure that the team understands the library requirements and at the same time is able to provide professional advices to the libraries.
3. The software vendor (or developer) should have long-term commitment on the further development of the software. Particularly:
   * The vendor should be quick to integrate emerging library standards and new technology to the software. These include: Web and Internet based access to the library catalog, XML and Dublin Core technology, Unicode, etc.

* The vendor must be specialized in library applications
* The vendor must have periodical upgrade release with new enhancements.

4. The library automation software must be able to support different kinds of library environment, these includes:
    * Bilingual interface and bilingual data
    * Sharing of cataloging records among the libraries
    * The software must run on computer and networking equipment commonly used in libraries
    * The initial purchase cost and the annual upgrade cost must be priced at a level affordable by various types of libraries
    * Asking if the version that is being demonstrated, for each module, is the same as the version in current release, i.e. what library would be getting if it brought the system. Also it should be asked if the modules demonstrated are included in the price quoted in the RFP.
    * Watching for cumbersome or awkward operations within or between functions.
    * Requesting that, in addition to the pre-planned demonstration scenarios, a few specific searches, operations, etc., be performed that are analogous to real situations in the library.
    * Taking note of specifics about the system's functionality that library staff judge to be particularly strong, as well as those that appear weak.

LIBRARY AUTOMATION CONSIDERATIONS

Library automation offers many opportunities to improve service to library users. Among other benefits, it makes easier for users to locate resources as well as allowing staff to better serve users by facilitating a multitude of staff tasks such as acquisitions, cataloguing, circulation and reference. On the other hand, the financial and staff commitment needed to move to an automated system or from one system to another is substantial and long-lasting. Automation cannot be regarded as a panacea for systemic problems in a library. Without the staff support and training no automation system can offer its full potential.

AUTOMATION COSTS

Automation involves different types of cost at various stages. Each stage and concerned cost has its own significance. While estimating the automation cost, all involved costs are to be considered.

* Planning and consulting costs
* Purchase of the system - hardware and software
* Purchase of network-specific hardware, software, and cabling
* Conversion of manual records into machine-readable form
* Ongoing operating costs
* Maintenance of system hardware and software

LIBRARY AUTOMATION TRENDS

Trends in library automation reflect the a going practices and the use of technology, directing the future applications. Following aspects clears the picture.

Integrated library system : A library automation system in which the various applications share one bibliographic database. Each system comes with a set of core modules as well as additional modules which can be added on if necessary.

Client-server architecture : Turnkey systems have become a thing of the past. A client/server system is identified by a more powerful server machine which handles database manipulation and retrieval while leaving the user interface to the desktop client software. This system shares the computational load between the client and server machines and provides the user a better experience through a faster interface.

Z39.50 : This is a protocol for computer-to-computer information retrieval. It allows users to access dissimilar library catalogues from the host institution's catalogue while using a familiar interface. Both a Z39.50 client and server are needed if library want to visit others and have others visit its catalogue. A Z39.50 client allows access to others'catalogues. A Z39.50 server allows others to access catalogue.

MARC and non-MARC compliance : First, library systems developed to use MARC records. Now systems are allowing for cataloguing formats, such as Internet resources.

Web-based patron catalogue : User access is greatly increased when catalogues can be accessed remotely via the World Wide Web. Better systems contain password-protection to allow users to access portions of their own records remotely.

UNICODE : This protocol expands the character set allowed and is essential for collections with materials in non-Roman languages. UNICODE encodes 65,000 different characters compared to the extended ASCII character set of 256 characters.

Core modules : Circulation, cataloguing and online public access catalogue are necessary minimums. Additional modules often cost extra. Additional modules include acquisitions, community information, course reserves, imaging, inter-library loan (ILL), materials booking and serials.

Digital library management : Separate products are being developed to handle digital library management.

The open source movement regarding digital library software is developing slowly within the library community as well. Products are still very locally-based and often just a single module, such as ILL, but as with all open source projects the potential for exponential growth without huge development budgets exists.

Automation Options

Various options are available for library automation. Library can choose the option based on its requirement.

* Acquiring software to run on a computer already in place
* Pursuing in-house software development
* Acquiring a turnkey system for the library
* Acquiring a turnkey system for a consortium of libraries
* Joining on existing bibliographic utility or service bureau
* Relying on the data-processing facilities and staff of the library's parent organization

Information gathering

While there are many vendors creating automation products, each library invariably finds their choices limited to a few when all of the variables are taken into account. A checklist of needs, wants and limitations should be compiled. Once a basic set of systems requirements has been established, vendors can be sent requests for proposals. However, most of the resources were created with North American and European libraries in mind. Each library may have its own particular checklist. The checklist should include an examination of module features such as which fields are indexed, as well as the long-term viability of the product and vendor. Library Systems includes a set of minimums to look at core modules (acquisitions, serials control, cataloguing, circulation and the user access catalog), installations, programmers and customer support ratio. Frequency of enhancements and the degree of customer input on these is worth mentioning as well.

Bibliographic Standards

"The creation of a high-quality, machine-readable database provides the cornerstone upon which all future automation efforts will rest. Vendors may come and go, hardware may become obsolete, software may be replaced, but a well-constructed, well-maintained database, with its accompanying local holdings, will be the library's transportable and viable link from system to system." — *extracted from John M. Cohn, etc. Planning for automation, 1992.*

Why bibliographic standards?

Bibliographic standards are needed and accepted :

* To maintain the portability of data
* To enable resource sharing

Overview of programs

In light of recent mergers and acquisitions, increasing development costs, it is important to select a software that belongs to a strong vendor and which that vendor is supporting well and with a future outlook. Despite this, a global outlook is growing among vendors as evidenced by UNICODE enhancements and as necessitated by the greater potential for growth. The common choice of Oracle as the database management system also reflects its position as the most-widely supported DBMS in the world.

Generally speaking, products can be divided into two groups. One group is based on multi-user operating systems and the other supports PC products. In general, the multi-user operating systems offer greater function and cater to academic as well as mid-size and large libraries. A library should seek to purchase a system from a vendor that has a significant number of its customers in the library's segment because product development will generally be focused on the needs of libraries that constitute the most important market segment for a vendor.

Numerous systems have been developed with the public and academic library markets in mind. Both the initial and maintenance costs for these systems are considerably cheaper. As well as targetting their modules to different client needs, however, these systems do not have as many functioning units. For example, reserves, interlibrary loan and journals modules are frequently missing. The customer support and development staffs of the smaller systems are often considerably smaller. The frequency, substance and responsiveness to customer requests of upgrades are superior among systems with a substantial development staff. Advanced searching capabilities are also limited.

A word should be said about two systems (TINLIB and Alice for Windows) relevant to the partner's experiences and past searches for automation systems, particularly since these systems have evolved and are not widely reviewed. Some facts about Softlink Alice (formerly called Alice for Windows) can be highlighted. It is Z39.50 server compliant, but not Z39.50 client compliant. Accessing other library catalogues can be problematic. It does not have an inter-library loan module either. In addition, its primary market was schools rather than academic institutions and its future enhancements will probably reflect that interest.

LIBRARY AUTOMATION AND LIBRARY STAFF

A lot is written about how to select a new library automation software, but not on how to efficiently and effectively implement the new system. Implementing a new automated library services is perhaps the biggest and most expensive task most libraries undertake. Since it affects the daily work flow of so many members of staff, good planning is essential for a smooth transition. Installing an automation process is not just a job for the automation department; all library employees must be involved if the automation is to be successful. For most libraries, the process of library automation is not familiar—that is, it's not something librarians do very often, so they don't have the proper structure in place. Most of the librarians don't know where to start, and plan lately.

INITIATING THE WORK AND PREPARING THE LIBRARY STAFF

Involving the library staff in the selection process of software for library automation is important for several reasons. First, if staff members were a part of the selection process, they will have geared up mentally for the task of automation, and hopefully, they will have allocated their work so as to set aside time for the process. They will also have a sense of ownership in the new system. The staff may not agree with the selection, but it will feel a part of the decision. And last, they will be more motivated to support with the implementation. The library staff members are the greatest allies of the librarians. When they feel they are a part of the library automation process, they may consider to help. New process will not improve every aspect of the existing system, so some staff members may end up with some cumbersome tasks. But if they are involved in the process up front, hopefully they will see the outcome of their efforts and will persevere. It is important to divide up the library automation tasks (looking at planning and policies, data inputting, and completing the modules) among the staff, but these need to be working groups of people. One person cannot and should not do everything. Most systems are too complex for one person to have all the expertise. It is important to make sure each working group gets broad representation from the entire library, and that good communication flows among members of the library staff.

To make the library automation process manageable, it is highly recommended hiring additional staff for the automation process. Besides the enormous task of the library automation process, there are still day-to-day activities to accomplish. The library can't stop buying, cataloging, or servicing the collections, yet the library has to add the huge load of moving data to a new system. Almost every library employee is affected by the automation. In addition to the actual automation, there will be clean-up small tasks such as rekeying data, catching up on backlog created during down time, verifying loads, training, creating new policies and procedures, and publicity and communication to the clientele.

SETTING A TIME FRAME

Overseeing an entire automation can be overwhelming, so it is necessary to break down the automation into smaller tasks and place them into a timeline with deadlines. Most process deadlines should only be viewed as a target and the anchor events are set and cannot easily be changed to different dates. Anchor events might include vendor training or cutting off the old system; other items in the timeline then revolve around these events. The anchor events provide leverage, both inside and outside the library, goals, successes, and stability throughout the entire automation. Knowing *what* has to be done *when* is essential, for library and the entire staff. Updating the timeline frequently, and sharing it is widely important.

TRAINING IS AN ESSENTIAL ASPECT

Some vendors offer training on the new system's operation at an additional cost. As much vendor training should be arranged as much library can afford, but because it can typically expensive, chances are that staff won't be able to afford enough. Usually, the training naturally concentrates more on the big picture and less on the detailed specifics of the system and how it works. The other result of the high cost of vendor training is that library can't afford to include as many staff as libraries like. Library staff be prepared to follow up any vendor training with in-house training.

DESIGNING GOOD TESTING STEPS

It is extremely helpful to have data prepared for testing and a checklist of potential problem items to look at during the implementation process. There will be numerous times in the automation process that the vendor will want library to test a section of library. If library doesn't have any examples in place beforehand, it won't have anything to compare it to. Is everything in the appropriate place? Did the vendor follow all of the requested specifications? It is essential to have specific examples that can be used to examine pre- and post- automated sections of library quickly and thoroughly. Staff members can also create problem records, which can be tested to assure that the vendor is dealing with incorrect data correctly. It's also a good idea to have a total number of records for each type of data library is migrating (bibliographic, authority, items, holdings, patron, vendor, open orders, fund, and outstanding circulation transactions) so library knows if the expected number of records actually migrates.

USER INVOLVEMENT

While automating the library, one important question arises whether faculty and students (or library users in general) be involved? There are many schools of thought on this one, and the answer probably depends greatly on the local environment. Typically, the faculty and students won't have a clue about the automation problems or solutions. However, involving constituents will help them understand how complex a automation really is. It will also get them on board, help spread the word, and may even help build library support. Involving faculty and students in the selection process can of course be the ideal. But, library should make sure they are trained, and involve them in the look and feel of the new system. Library's own staff will provide the most useful suggestions, so users can be used as a focus group.

THE NEW AUTOMATION SYSTEM

No doubt, library will not be able to get everything done by the date user wishes. In addition, there are some things that library can make more informed decisions about after staff fully understand the implications of the decision. New services are probably why library bought the system that library did, but do they need to be activated right away? Could things like self check-out, user placed services, and cataloging of images and sound clips be postponed slightly? It is important to set realistic priorities and communicate that other features will have to wait until library gets to them in post-production.

When does library close down the conventional system? This is a real balancing act. On one hand, library will be paying maintenance on two systems during any overlap period. This may be something library can't afford. Yet, there is usually some information that can't be implemented and needs to be retained long enough to see another process through or until it can be automated. It is a belief is that it is cheaper to hire temporary staff to rekey data into the new system or to save data library know will not automate in paper or electronic form than it is to keep the old system active until existing staff has time to deal with non-implemented tasks. Library might consider phasing out the conventional system by canceling aspects of the software, maintenance, or operating system while retaining only the aspects library still needs.

PHASING OUT

Regarding library automation some things can be done ahead of time. If library needs higher levels of computers or particular software, it is necessary to get them installed before the automation if library can. It is also necessary to prepare the equipment room with any required electrical, data and start examining policies in circulation. A new system will not have all of the same functionality and options. Depending on what is available in the new automated library system, policies, will most likely have to change. Anything the library can do ahead of time will make vital that much easier during the automation.

VENDOR COMMUNICATION

Keeping the vendor in line can be difficult. But it is found that vendors are allies of library, so they can be used for advantage. They are used to support the library automation and they probably have experience in automating libraries from the current system. However, library must keep on top of them. It is important to remember that the vendor is most likely working on multiple installations at the same time; they often won't have time to pursue the questions fully, unless they are asked the question several times. It is necessary to be fair with the vendor; without expecting the world, but staying firm on what library's needs are. After automation, it is essential to re-read and verify that all vendor promises.

It is important to resolve as many outstanding problems and transactions in the existing conventional system before the initiation of library automation. In addition, it is essential to perform as many purges as possible to get rid of unwanted user records and logically deleted but not purged bibliographic, holding, and item records. In every automation there may be strange phenomena with outstanding transactions. They've never been serious, but a little work up front can save library tenfold later. Library must know its priorities. If the library doesn't know current priorities and practices, it may lead towards trouble. Library can usually ask the vendor to change blocks of records during the automation.

It is important to understand the amount of technical work that will occur during automation. However, it is usually forgotten the equally important aspect of organizing the project so that the automation is accomplished in an effective and efficient manner. Yet, this alone does not involve enough staff members with expertise. A successful task can only be achieved by bringing the entire library staff together so that the strengths of each employee are maximized in such a way that the automation process is not bogged down. Most important of all is communication—communication of what each person's role is in the automation; what is going on; and when specific tasks, decisions, and training are to be accomplished.

IMPORTANT DATA AND INPUTTING ESSENTIALS

Following items are essential to consider important while library automation:

* Bibliographic records for each location
* Periodical payment data, receipt notes, check-ins, and action dates
* Acquisition payment data, receipt notes, and action dates
* Long bibliographic records

* Long bibliographic fields such as 520 and 505 tags
* Bibliographic records with several copies in various locations
* Bibliographic records with many attached items
* Long authority records
* All MARC formats (videos, CDs, microfiche, music ...)
* Patron records (status, active, address, fines, active transactions)
* Call numbers with semicolons, or of atypical origin
* Order records
* Claim notes and action dates
* Open orders by type (continuation ...) and prices
* 949 fields of bibliographic record
* Short MARC records
* Copy holding notes and 852, 866, 867, 868, and 899 fields
* Long and/or complex call numbers (government documents, Dewey, Cutter)

BENEFITS TO LIBRARIANS

Librarian is the front line for library automation and success and he or she requires library automation software that makes library's life easier, not harder. Automation software will enable librarian to:

* Prove his value to administrators. Software gives the tools, such as vocabulary capabilities, to actually boost users achievement in a tangible way.
* Put professional skills to better use. Software's well-known ease-of-use will help him switch gears from putting out software fires to spending quality time with the users.

BENEFITS TO IT MANAGERS

IT managers are tasked to do more with less, and they need library automation software that can answer. Library software may enable to:

* Relieve the support load of the staff.
* Minimize the resource drain of multiple servers at multiple sites. Software gives the ability to update/ upgrade and backup the entire district from one primary location.
* Eliminate upgrade and backup hassles. Automation software's update, upgrades, and critical data backups are all done automatically within the software.

Benefits to Administrators

Library automation software allows to efficiently managing varied library education and management responsibilities with the following unique capabilities:

* Improve reading scores.
* Universal access.
* Relieve the resource drain of extra support staff just to manage library automation software. Software's automatic updates, technical support, and renowned ease-of-use allow re-deploying valuable resources to more important areas.

Reporting problems

Initial and on-going training will be needed at both the staff and user levels. As well, such a group can provide a stronger voice to enhancement suggestions to the vendors. A complex and exciting decision waits. As more information is received from the partners about their needs and desires, demonstrations by vendors and bibliographic utilities can be arranged for later on.

REFERENCE

1. Planning for automation : a how-to-do-it manual for librarians /John M. Cohn, Ann L. Kelsey, Keith Michael Fiels. 2nd ed. New York : Neal-Schuman Publishers, Inc., 1997
2. Breeding, Marshall. "A new look at large-scale automation systems," *Computers in Libraries*. 19:8 (Sept 1999): 36-40.
3. *Library Systems Newsletter*. ALA TechSource. (Jan 2000-Dec 2001)

4. Nkereuwem, Edet E. "Theory and practice of cataloguing library materials by using TINLIB software," *International Cataloguing and Bibliographic Control.* 26:3 (July/Sept 1997): 66-67.
5. libinfo.com/libauto.html
6. www.latcorp.com
7. www.jadavpur.edu/library/library_automation.htm
8. libraryautomation.com
9. www.libraryhq.com/automation.html
10. en.wikipedia.org/wiki/Integrated_library_system
11. www.cenlib.iitm.ac.in/docs/library/index.php?page=Library_Automation
12. www.librarytechnology.org
13. Northover, Steven A. "A checklist for choice of an OPAC," *Library Automated System Information Exchange* 21:2 (Sept/Oct 1990): 51-56.
14. Saffady, William. "The status of library automation," *Library Technology Reports* (Jan-Feb 2000).
15. Saffady, William. *Introduction to automation for librarians.* 4th ed. Chicago: American Library Association, 1999.
16. www.bitpipe.com/tlist/Library-Automation-Software.html
17. igmlnet.uohyd.ernet.in:8000/automation.html
18. www.phpmylibrary.org
19. www.intota.com/multisearch.asp
20. www.rfid4u.com/downloads/Library%20Automation% 20Using %20RFID.pdf
21. www.librarytechnology.org/vend-search.pl
22. www.fcla.edu
23. www.infosciencetoday.org/automation.htm
24. www.easylibsoft.com
25. Carter, Nancy and Scott Seaman. "Circulation Policy Issues and System Automation: You Can't Beat a Move for Cleaning House." *Colorado Libraries*, Winter 1997, pp. 32–35.
26. Cleyle, Susan. "Avoiding the Implementation Blues." *Feliciter*, Vol. 42, 1996, pp. 32–36.
27. Copland, Nora S., James F. Farmer, and Patricia A. Smith. "Data Automation: A Brief Primer." *Colorado Libraries*, Winter 1997, pp. 22–24.
28. publib.boulder.ibm.com/infocenter/pcomhelp/v5r9/topic/com.ibm.pcomm.doc/books/html/host_access08.htm
29. www.eosintl.com
30. ieeexplore.ieee.org/iel5/85/21773/01010068.pdf

31. www.booksys.com
32. www.libsuite.com/library_auto.htm
33. www.bitpipe.com/tlist/Library-Automation-Software.html
34. 3d2f.com/programs/2-392-biblio-school-download.shtml
35. www.libwin.com
36. Culbertson, Michael. "The Training Aspects of System Automation." *Colorado Libraries*, Winter 1997, pp. 28–30.
37. Garrison, William A. "Issues in System Implementation." *Colorado Libraries*, Winter 1997, pp. 13–16.
38. www.librarytechnology.org/
39. www.libsuite.com/library_auto.htm
40. http://www.igidr.ac.in/lib/paper1.htm
41. en.wikipedia.org/wiki/Integrated_library_system

6
Library Automation Software

SOUL

General Features

University libraries are complex entities, having large collections and serving a huge clientele. To carry out various operations in a library effectively, there is a need for automation. Computer and communication Technologies have brought revolutionary changes in the information acquisition, processing, storage, retrieval and dissemination. Keeping in view the latest trends in Information Technology (IT), INFLIBNET Center has developed a Windows based Library Management Software **"SOUL"**, which provides total solution for Library Automation. SOUL is designed using Client-Server Architecture which imparts extra strength to storage capacity, multiple access to single database, various levels of security, back up and storage facilities etc. This software has been designed after a comprehensive study of different library related functions practiced in university libraries. It has MS-SQL Server 7.0 or higher as the back end. This user friendly software is quite easy to work with. SOUL handles Indian languages/scripts using ISM Publisher of C-DAC. There is an effort is going on to develop a new version of software based on MARC 21 and Unicode standards and RFID protocols for electronic surveillance. The software comprises following modules.

* Acquisition
* Catalogue
* Circulation
* OPAC
* Serial Control
* Administration

The in-built network feature of the software will allow multiple libraries of the same university to function together as well as access to the distributed databases installed at various university libraries and union catalogue mounted at INFLIBNET using VSAT network.

ACQUISITION

The Acquisition module enables library staff to handle the following major functions related to acquisition of library material.

* Suggestions management.
* Ordering, cancellation and reminders.
* Receiving.
* Payment including fund control.
* Master file management such as currency, budget, vendors, publishers etc.

Through this module library staff can search the entire database of library holdings for the purpose of duplicate checking etc. Using various combinations, number of reports could be generated.

CATALOGUE

This module is used for retrospective conversion of books, technical processing of books received from Acquisition Section, printing a range of records for verification, searching by title and accession numbers, authority files for publishers etc. One of the unique features of SOUL is accessing to authority files on screen as well as selection box in various fields of records. Once the proposed VSAT Network by INFLIBNET is commissioned this module will take care of automatic replication of data into union catalogue. This will avoid exchange of library data on physical media i.e. floppy, tape etc. This module covers following functions.

* Cataloguing Process.
* Catalogue Search.
* User Services.
* Catalogue card generation.
* Authority file maintenance.
* Retro conversion.
* Stock verification.
* Report Generation.
* Union database maintenance (local level).
* Export/Import of records.

CIRCULATION

This module will take care of all possible functions deal in a university library setup. Starting from membership management, maintenance of status of library items can be handled using SOUL. The circulation transactions viz. issue, return, renewal, reserve, recall, hold can be successfully done. The complex functions like fine management for each category of user and material-wise can also be effectively managed. Inter library loan, searching the status of every member or library item is possible. Reminders for overdue material, generation of various reports have also been covered.

* Membership
* Transaction
* Inter-Library Loan
* Over due Charges
* Reminder

* Search Status
* Report Generation
* Maintanance - Lost, Binding, Damaged, Missing, Withdrawal, Reports

ON-LINE PUBLIC ACCESS CATALOGUE (OPAC)

The On-line Public Access Catalogue (OPAC) of SOUL is a window to the library collection. Using the user friendly menus, user can search for an item available in the library by author, title, corporate author, conference name, subject descriptors, class number etc. The Boolean search enables the user to conduct the search using all combinations including type of material, language, year etc. Besides, this will serve as first point of information about the library and also gateway for accessing INTERNET, union databases, or any other external sources. Web OPAC facility is also there.

* Books/Theses
* Serials
* Member Status
* Other Resources
* INFLIBNET Access

SERIAL CONTROL

The complex job of keeping track of serials can easily and effectively be handled using SOUL through its Serial Control module. This module broadly handles following functions.

* Suggestions.
* Subscription (renewal and new subscription).
* Payment including fund control etc.
* Check in of issues including prediction of issues arrival.
* Reminder generation.
* Binding management.
* Search status of every item.
* Master database management.

* Reports generation etc.

ADMINISTRATION

User administration facility in this module is used for creating new users and giving them right for accessing different modules.

Benefits of using SOUL

* Available at nominal cost to university libraries.
* Software designed and developed exclusively to work under university environment.
* Network feature of the software will allow multiple libraries of university to function together.
* Exhaustive training at INFLIBNET supported by comprehensive manual.
* On-site training.
* Free updation/modification.
* Free technical assistance.
* Follows INFLIBNET's recommended standards and formats such as CCF, MARC 21, AACR2, LCSH.

SLIM

SLIM library management software is a product of Algorhythms Consultants Pvt. Ltd. Pune, established in 1989. The software is constantly upgraded to remain in the forefront of technological developments in the field of IT and to cater to the changing scenarios in managing libraries. SLIM21 is an integrated, multi-user, multi-tasking library information software for the Windows environment.

SLIM21 helps to catalogue books, films, sound recordings, drawings, clippings, articles, reports, letters, pamphlets, serials publications... all those things that contain information so vital to the organization. SLIM21 cataloguing adheres to popular international standards. This means that one can exchange data with the world. Retrieval of the data is simple, fast and efficient. Even a catchy phrase in the description of the catalogued item can be used for searching. SLIM21 is designed and developed in **modules** to take care of complete functionality required for automating libraries. Library can configure SLIM21 software for library by selecting one or more of these standard and add-on modules. These modules exhibit **features** that make SLIM21 a top class software.

SLIM21 modules work on the same data from different nodes of a network. They can be installed independent of each other on different desktops. SLIM21 works just as well on a stand-alone machine as it does in a network of Computers. Moreover, library can be browsed through the Internet / intranet with SLIM21.

The greatest advantage of SLIM21 is that library can make data entries in any language library wants since it supports **Unicode**. Apart from SLIM, SLIM++ is a 32-bit application developed on the Windows platform. This software comes in various modules and each one of them can get integrated with the other modules at any given point of time. This software comes with an unlimited multi user license. The latest product - SLIM21 - a product on Microsoft dot net technology offers remarkable features to serve the requirements of a digital library. SLIM21 provides for inclusion or exclusion of metadata elements that are used in the creation of the catalogue entry. SLIM21 also stores records of such data entry formats in its database. The cataloguing of digital items is done using exactly what the library chooses. SLIM21 provides initial versions of data entry formats for various types of items. Libraries can build their own formats by modifying these. SLIM21 covers, and extends if required, the popular Dublin Core format. SLIM21 provides for data entry and storage in Unicode. Use of the Unicode technology makes it possible to install as many scripts as required on the computer and use them in every interaction with the computer. SLIM21 is a must for libraries with collection in Indian Languages.

BASIC MODULES

* Cataloguing
* Circulation
* Serials Control
* Acquisition
* OPAC

Enterprise Modules

* Web based OPAC
* Current Awareness Service (CAS)
* Selective Dissemination Information (SDI)
* Web Proposals
* Inter Library Loan
* Statistical Analysis

L2L Modules (Library to library resource sharing

* Import Export
* Z39.50 Server
* Z39.50 Client

Additional utilities

* Colon classification shelving order
* Smart Card / RFID interface
* Touch Chip Interface (Biometrics)
* Library Map
* Newspaper monthly billing
* News clipping publishing

LiBSUITE

LiBSUITE is a web enabled library automation software which can be used for automating various libraries of any size and having any specific requirements.

Libraries are of different types and have specific needs in their own areas of expertise. LiBSUITE is an integrated package which provides an end to end solution for different types of Libraries.It is one of its kind Enterprise level library automation software which is used by many big corporate to manage their collections at various locations. It is available in multi-lingual format supporting various vernacular languages.

LiBSUITE is used by more than five hundred libraries all across the globe. Following are various versions of Libsuite. These versions are made specifically to cater to different types oof industries or segments. Pl. click on details to get more details about any specific version.

Modules

LiBSUITE has various modules catering to specific needs of every department. Cataloguing, Circulation, Acquisition, Serial Control, Budgeting etc. are the main modules of LiBSUITE. Apart from this there are many more modules like Stock taking, Bindery Control, Software Product Cataloguing and many more such modules are available.

* Cataloguing
* Circulation
* Serial Control
* Budget Control
* Acquisition

LiBSUITE is a Bar Code and Smart Card enabled software.

Toolbars facility : To directly navigate into the functional feature within a module. The frequently used features have toolbars which can directly take the user to that function.

Tool tip text option : All major options on the screen have a tool tip text associated with it explaining the functionality in detail.

Context based On-line help : Complete Context based On-line help has been provided to the User while working in any module or using any feature of the software.

Tool tip text option : All major options on the screen have a tool tip text associated with it explaining the functionality in detail.

	Product Code	*Name*	*Segment*	*Technology*	*Backend Supported*
1.	SALIB0001	Libsuite Personal	For small software for a single user and a very tiny library.	Client Server	MS ACCESS
2.	SALIB0002	Libsuite Educational	For Educational Institutes hence very effective Circulation Module. Generate all NAAC related reports.	Client Server	MS ACCESS
3.	SALIB0003	Libsuite Standard	For a medium Library but no provision of choosing the backend.	Client Server	MS ACCESS
4.	SALIB0004	Libsuite Professional	For a medium Library with possibility of choosing various backends.	Client Server	MS ACCESS SQL 7, Oracle 8i
5.	SALIB0005	Libsuite Enterprise (Web Enabled)	For Large Libraries with big Data but there will be only one server license.	Web Enabled / Server Centric	MS ACCESS SQL 7, Oracle 8i
6.	SALIB0006	Libsuite Multilocation Upto 5 Locations (Web Enabled)	For Libraries having multiple locations but data stored on the same server	Web Enabled / Server Centric	MS ACCESS SQL 7, Oracle 8i

7.	SALIB0007	Libsuite Multilingual	Library can store information not just from English but any language	Web Enabled / Server Centric	MS ACCESS SQL 7, Oracle 8i
8.	SALIB0008	Web OPAC (Web Enabled)	For users to access the library information	Web Enabled / Server Centric	MS ACCESS SQL 7, Oracle 8i
9.	SALIB0009	LiBONWEB	Utility to make the data available on Web	Web Enabled / Server Centric	MS ACCESS SQL 7, Oracle 8i
10.	SALIB0010	DocuMantra	Maintaining the documentation in digital format	Web Enabled / Server Centric	MS ACCESS SQL 7, Oracle 8i
11.	SALIB0011	LiBSUITE RFID	Radio Frequency Identification Systems	Client Server	MS ACCESS SQL 7, Oracle 8i

Context based On-line help : Complete Context based On-line help has been provided to the User while working in any module or using any feature of the software.

Multilingual Compatibility : The package is compatible to multilingual fonts. The data content or information can also contain a mix of different languages at one time.

Desktop background selection by user : The on screen background image of Librarian is selectable by the user who would like to view it throughout the session.

Security Features : Data security is of prime importance in a networked environment and as such every Menu option has rights definable as per Users profile. The Login and Passwords may be assigned to groups of individuals depending on their designation or department. In addition, every User can have an individual password to access his personal information (e.g. Books outstanding etc.). The Members and Operators Photograph and Signature can be provided and viewed online during Front Desk Operations

ALICE

Alice efficiently and effectively assists library in the management and control of the library. It performs all the functions library require in a library system. Alice is reliable and proven product. Over the last twenty five years it has grown and achieved the status of leading library automation software of the world.

Salient Features

* Electronic Documents Cataloguing
* Boolean Multiple Fields Search
* Reminder Notice Via e-mail to members
* Message popping for overdue of Resource
* Bulletin Board Failed Search Records Automatic Keyword Generation
* See and See Also Reference Library Map

* Multimedia Live Demonstration Member Card Generation With photograph
* Barcode Labels and Spine Labels from software database Printing by using laser printer (no need to purchase barcode printer). Random barcode generation is also possible
* Reservation and booking of Resources in Intranet/ Internet Environment

ALICE Modules

* Acquisitions
* Internet Suite
* Journal Indexing
* Multilingual
* Multimedia
* Periodicals
* Quick Pic
* Rapid Retro
* SDI
* Secure Web Browser
* Text Book Control
* Union Catalogue
* Web Cataloguer
* Z39.50Server

LIBERTY3

Salient Features

* **Fully integrated :** Liberty3 is a fully web-based and completely integrated solution.
* **Ease of use :** Liberty3 is intuitive, has a consistent user interface and incorporates on-screen help.
* **Latest technology :** Liberty3 is built on world leading technology which facilitates integration with other systems, offers high-level security and fast response times.

Liberty3 incorporates extensive data validation throughout the system to ensure absolute accuracy of information.

* **Flexible and Customisable :** Liberty3 offers a wealth of functionality and usability straight out of the box. It also offers an unprecedented level of customisability to meet unique library requirements.
* **Resource Management :** Liberty3 has in-built functionality to manage and schedule all library resources and assets.
* **Accessible :** Liberty3 provides 24 by 7 "access anywhere" availability for library staff and library users.
* **Affordable :** Liberty3 is attractively priced and offers outstanding value for money.
* **Minimal Support from IT Staff :** Liberty3 is designed to require minimal support from IT staff and maximum flexibility for library staff to deliver timely and quality services to clients.
* **Design :** Liberty3 is designed by both library and information specialists.

LIBERTY3 Modules

* Aquisitions
* Cataloguing
* Circulation
* Borrowers
* Serials
* Inter Loans
* System
* Customisable Reports

Purpose of Acquisitions Module

* Provides a quick catalogue / interloan search at the new order entry screen .
* Foreign currency calculator with exchange rate reminders.

* Allocate committed and spent amounts to various cost centres and fund combinations .
* Report the committed and spent amounts by cost centre or fund, for the year-to-date or any particular month.
* Allows partial arrival, partial payments and cancellations.Order letters may also be printed or e-mailed.

Purpose of Cataloguing

* Full text searching throughout Individual copy management, Thesaurus searching and control for authority fields
* Define the own bibliographic types and adjust the input form to suit
* Includes a number of standard bibliographic reports
* Copy records from BookWhere directly into the catalogue
* Provide a z39.50 server interface to publish the catalogue to the world
* Web based OPAC with borrower self-help functionality

Purpose of Circulation Module

* Simple issue and return processing system
* Optional plug in receipt printer at the circulation desk
* 'First-available' reservation processing and notification
* Self-issue and self-return workstation option
* Offline data capture option
* Support for mobile libraries

Purpose of Borrowers Module

* On-the-fly management options from the circulation station
* Auto-loading of borrower details from an external source
* Self-service functions available through the web OPAC
* Support for notices to be e-mailed to the borrower
* Alerting service available for new topical catalogue entries and journal issue arrival

Purpose of Serials Module

* Automated check-in processing to support a publisher's barcode
* Flexible issue prediction system supports many varied types of journal schedules
* Claim management, automatically identifies missing issues and broken subscriptions
* Subscription management ties into the acquisition budget report
* Record contents page information or add brief article records

Purpose of Interloans Module

* Record and manage interloans requested
* Record and manage interloans supplied
* Non-supply and cancellation processing
* Interloan charges tie into the acquisition budget report
* History of all interloan requests maintained online

Purpose of System Module

* A secure login process and application level security
* Individual user preferences
* Customisable interface
* Customisable forms and databases

Purpose of Customisable Reports Module

* Report writer generates output direct to Microsoft Word
* Global editing utility
* Ability to modify authority lists; branches, formats, collections, item status, categories, etc

OLIVER LIBRARY AUTOMATION SOFTWARE

For School Group Libraries

Softlink develops and supports leading library management solutions for small to large school libraries. Softlink has been partnering with Australian schools for over 21 years and through these relationships has built a thorough understanding of the unique needs and requirements of school libraries. Oliver is the latest addition to the suite of library management solutions. Oliver is a fully integrated web based library management solution, suitable for all schools in either single or multi campus libraries.

This understanding has lead to the development of a range of library management solutions that are being used by thousands of school.

Oliver - Versatile Web-based Library and Information Management System for Schools

Oliver is a web-based library and information management system suitable for all schools in either single or multi-campus libraries. All functions are fully integrated. Oliver offers full-text searching for fast, efficient and powerful retrieval.

Running on Windows-based web server, Oliver offers the following features:

* All functions are fully integrated
* Suitable for either single or multi-campus libraries
* All functions are performed through a standard web browser
* Fast keyword searching over every field in every record
* Flexible and accurate searching from basic to advanced
* Boolean and proximity operators
* Live SCIS downloads
* Ability to customise fonts, colours, data displays and input forms
* Reports exported directly to MS Word or Excel

Oliver also supports multiple OPACs. Different student populations can have different OPACs to meet their needs. Softlink also offer Oliver ASP - all of the power of Oliver, without the installation, hardware, or IT support requirements. Library can start using Oliver immediately, connecting directly to our data center over the web.

Important features

* Easy to use for both students and teachers
* Fully web-based
* Quick to learn for library and teaching staff
* Suitable for single and multi-campus libraries
* All functions are fully integrated in a consistent user interface
* Keyword searching over every field in every record
* Keyboard shortcuts for efficient operation
* Live SCIS downloads
* Customisable font, colours, data displays and input form
* Reports directly to MS Word or Excel

LIBRARIAN - LIBRARY MANAGEMENT SOFTWARE

Librarian is an advanced and elegant library management software which is designed and supported by Growth, India. Suitable for special, medical and legal libraries, colleges, schools, charities and other resource centres, Librarian software has revolutionized library automation in various institutions. Over 95% of the sites that have ever purchased Librarian are still using it.

Librarian Software is a complete Library Management Software capable to manage all the functionalities of library and dissemination of information at the right time to the right people. Suitable for small to big libraries including medical and legal libraries, colleges, schools, universities, corporate houses, charitable trust and other academic resource centers. Other then books *LIBRARIAN* Software can be used to manage CDs, DVDs, Flip Charts, Drawings, Scan Documents, X-Ray Reports and any other present and future medias..

LIBRARIAN Software can also be used by the Music / Movie Libraries and associations to manage stock and circulation of their collections.

LIBRARIAN is now available in following editions :-

* LIBRARIAN - School Edition
* LIBRARIAN - Nano Edition
* LIBRARIAN - Professional Edition
* LIBRARIAN - Enterprise Edition

All flavors of LIBRARIAN support Unicode and other internationally known standards

Salient Features

* **WebOPAC** for online search of information
* Supports **Unicode** (Multilingual)
* Supports Multi-Users Environment and Multi-Location Web Interface
* Web Interface : Access to free e-journals and e-books through the **software**
* Multimedia Resource **Management** (Embedded)
* Internet Resource **Management**
* **LIBRARIAN Software** uses **MARC** format for data input
* **LIBRARIAN Software** has user interactive module for search called Boolean Search
* Contains more then 20 fields for search
* Link Menu facility i.e. **LIBRARIAN Software** links with other applications like MS-WORD.
* Circulation Module with member/borrower profile and lending rules
* Budget **Management** Module handles total **library** budgeting
* Cataloguing as per **AACR2** formats on Author, Title, Subject, Series
* SDI Service
* Object-Integration like Audio/Video Cassette, CD, Flip Charts, Posters can be entered and can be searched through keywords
* In-build BARCODE **software** and printing

* Around 50 types of reports available, which covers all the reporting part of a **library** administration and **management**

Minimum System Requirement

Single User

* Pentium based Processor (Celeron or higher)
* 128 MB RAM or more
* 2 GB or more Hard Drive
* CDROM Drive for Installation purpose only
* Color SVGA Monitor or higher
* MS Windows 2K/XP

Multi User

Server:

* Pentium based Processor (Celeron or higher)
* 256 MB RAM or more
* 4 GB or more Hard Drive
* CDROM Drive for Installation purpose only
* Color SVGA Monitor or higher
* MS Windows 2K Server
* Database Server (in case of MS SQL / Oracle)

Client

* Pentium based Processor (Celeron or higher)
* 128 MB RAM or more
* Local Hard Drive
* Color SVGA Monitor or higher
* MS Windows 2K/X

NETTCLIPS

Information source for a library need not be restricted to books / journals but could be anything like a newspaper clip, a manuscript or an extract abstract of some other documents. The information could be simple text or picture or a combination. The document if it is something like a manuscript could be too fragile to be handled and used frequently by many people. The NETTCLIPS can convert such documents and many more into digital format. While the original could be safely preserved, the e-document could be made available to authorised users. In addition to the contents, the e-document will provide the look and feel of the original document to the user.

NETTCLIPS is designed especially for such application. Using NETTCLIPS library can scan the document in one part or several parts, catalogue the scanned image and store in a compressed format. Retrieval, handling of image like zooming, rotating, printing is all a matter of routine.

NETTCLIPS integrates seamlessly with the library automation software Nettlib. With this facility library can scan important portions of a bibliographic material like index ages of a book etc, and save as an integral part of the bibliographic record. Thus, users could not only see the title and authors of a record but also read the scanned abstracts, index, and contents.

NETTCLIPS can also independently handle catalogued images like newspaper clips of its interest.

Typical Applications

* Cataloguing, saving, retrieving and printing newspaper clips etc.
* Archiving important documents in electro optical media.
* Archiving office orders, circulars etc of statutory nature
* Putting documents in public domain, making it there available over the net

BARCODE/CIRCULATION READERS

Linear Imager

The linear imager is a high performance bar code scanner, using the latest technology to set a new standard for hand held scanners. It combines a bright, sharp aiming line with high resolution imaging for exceptional reading performance.

CCD Wedge Scanner

Designed to work with Alexandria, special LED technology enables the CCD Wedge scanner to read barcodes without touching the label.

Laser Scanner

The Laser Scanner with stand is a high-performance and pre-configured. Barcodes scanned by the Laser Scanner read quickly and accurately.

Omnidirectional Scanner

The COMPanion Omnidirectional Scanner is a high-performance barcode reader that offers the convenience of a multiple/omnidirectional scan pattern for quick and easy bar code label reading.

PORTABLE READERS

Wireless Linear Imaging Scanner

Wireless Linear Imaging Scanner is a high performance barcode reader that can be used for both inventory and circulation.

Palm Scanner

Palm III Laser Scanner offers an easy, ergonomic pocket design, allowing media personnel to perform multiple tasks while doing inventory.

Slip Printer

Keep students/patrons informed of their checkout due dates, fines, holds/reservations, and other transactions with Textbook Tracker's Slip Printer (receipt/transaction printer).

Photocomposed Bar Code Labels

These high-quality long-lasting labels are made using a photographic process that makes the barcode image an integral part of the label. They come with high-quality lamination and adhesives to increase durability.

P/N V0032 - Spine Labels, 4,800/pkg, 1 x 1-5/8, 8 labels across by 6

>labels down per sheet.

>P/N V0034 - Spine Labels, 5,000/pkg, 1-1/2 x 1, 5 labels across by 10

>labels down per sheet.

Laser Printed Bar Code Labels

These standard labels are made using a printing process that laces the ink on top of the label.

REFERENCES

1. LAD - Library Automation Division *http://www.informindia.co.in/Divisions/LAD.htm*
2. *www.biblio-tech.com*
3. libinfo.com/vendors-systems.html
4. *www.linux-directory.com/filesystems/Design-of-Library-Automation-Systems-File-Structures-Data-Structures*
5. *www.fsc.follett.com*
6. www.researchandmarkets.com/reports/450620/library_automation_systems_and_services_global
7. *www.librarysoftware.in*
8. *www.business.com/.../software_applications/*
9. *www.rightonlibrarysoftware.com*
10. libraryautomation.com
11. *www.l4u.com/*
12. *www.surpasssoftware.com*
13. *www.silas.org.sg*
14. *www.growthcompusoft.com/html/sw/plibrary.html*
15. *www.resourcemate.com*
16. *www.fsc.follett.com*
17. *www.mlasolutions.com*
18. *www.goalexandria.com*
19. www.softlink.co.uk
20. *www.librarysoftware.in*
21. *http://bg.computersinlibraries.com/*
22. http://www.goalexandria.com/solutions/accessories/index.html

7
Digital Library

Digital technology is revolutionizing the traditional concepts of preservation and access in library and archive communities. Although traditional preservation methods have ensured the longevity of endangered research materials, it has sometime been at the cost of reduced access. With digital technology, images are used to reproduce rare items, allowing for virtually universal copying, distribution and access. The technology also makes it possible to bring collections of disparate holdings together in digital form, making resource sharing more feasible. Digital technology is a driving force in many of the changes occurring in higher education. In response to the specialized needs of our users, the research library sector has led the way, seeking to create and use information resources in the digital environment. Others may be weighing the implications for library management in the implementation of digital technologies. The challenge in utilizing digital technology for enhanced access is the need to secure that investment and guarantee long-term access to that information. Making the right information available to the right person, at the right time is the aim of any modern library service today. Digital library is the electronic library in which information are stored in the digital from organized transmitted and displayed without any intervening conversion process with the availability of computer capable of computing at high speed and having large disc storage space, it is possible to store information in form of high quality graphics coloured images voice signals and video clips at very affordable coast. Digital library does not mean library in classical sense. It is a group of distributed repositories that users see as a single repository in a digital form. Digital library is a logical extension of the networked environment and the development triggered thereof and provide the users with coherent access to a very rare, organized repository of information and knowledge. In a sense it is a global virtual library, the library of thousands of networked electronic libraries/ databases.

WHAT IS DIGITAL LIBRARY?

Due to the advances in information technology and the popularity of the Internet, more and more information resources, which were once available only in books and journals, are now available electronically on the network. Libraries are no longer bound within their walls. Not only the library has the option to access a wide range of databases, but also the alternative to digitize their resources and mount them on the network to provide broader access of its collection.

Synonyms:

A digital library can also be known as :

* Library Without Walls
* Networked Library
* Virtual Library
* Electronic Library
* Digital Library

A library can be considered as a digital library if it provides:

* access to digital information by using a variety of networks, including the Internet
* services in an automated environment

Typically, a digital library usually has:

* Library automation system
* Web server as gateway to digital resources
* Subscriptions to various web-based resources
* CD-ROM network
* Electronic document delivery
* Collections of electronic journals and electronic books
* Digital libraries projects
* Internet resources selection etc.

Two other meanings of digital libraries are as following:

* Digital Libraries, as a discipline, refers to the researches on the theories and technologies for the building digital libraries.
* Digital Library Projects refer to the digitization of library materials for access through the network.

In the context of digital library four assumptions can be identified

(a) Digital library is totally a paperless system with computer network and communication facilities.

(b) It is suppose monetary facility to traditional library for faster retrieval and better communication.

(c) Any individual can have a direct link to electronic publishers their agents/through computer communication facilities with or without knowledge of libraranship.

(d) Any communication interested organization or an individual with of without knowledge of librarianship having direct access to an electronic publisher can become local promotional trade agent in information service like one of the IST/STD centers electronic typing center reprography centers etc.

A WORKING DEFINITION OF DIGITAL LIBRARY

For Digital Library Foundation (DLF) partners effectively to "federate" digital libraries, they share a common understanding of what a digital library is. Their working definition is as following:

"Digital libraries are organizations that provide the resources, including the specialized staff, to select, structure, offer intellectual access to, interpret, distribute, preserve the integrity of, and ensure the persistence over time of collections of digital works so that they are readily and economically available for use by a defined community or set of communities."

Definitely, the concept of digital library has multiple senses that one might invoke in various contexts. For example, the concept may refer simply to the notion of collection without reference to organization, intellectual accessibility or service attributes. This extended sense seems to be in play, for example, when we hear the World Wide Web described as a digital library. The concept might also refer to the organization underlying the collection, or even more specifically to the computer-based system in which the collection resides. Digital *libraries* might also be distinguished from digital *archives* to emphasize the nature of the works collected to stress the preservation function. Digital libraries extend and integrate approaches adopted in traditional libraries, as well as in distributed information systems, to yield high-end information systems, services, and institutions. Digital libraries are distinguished in that they afford services connected with each of the phases of the lifecycle. They integrate technologies from a variety of disciplines to help realize the designs articulated by early visionaries. Since the early 1990s, the digital library field has emerged as an important area of research and development. There are hundreds of projects and thousands of reports/publications describing them.

The scope and definition of the field of digital libraries has been the subject of intensive debate. Here some combinations are displaying the true nature of digital libraries:

* Library++ = library + archive + museum + ?
* Distributed information system + organization + effective interfaces

* User community + collection (content) + services

Thus, digital libraries are complex systems that:

1. help satisfy information needs of users,
2. locate and present information in usable ways,
3. organize information in usable ways, and
4. communicate information with users and computers.

In the perspectives of library and information services, we can say that the Digital Library is:

1. Organized collection of multimedia and other types of resources.
2. Resources are available in computer process able form.
3. The function of acquisition, storage, preservation, retrieval is carried out through the use of digital technology.
4. Access to the entire collection is globally available.
5. Support users in dealing with information objects

MANIFESTATION OF DIGITAL LIBRARY

Integrated aspects of digital Library: digital library can be manifested into following number of purposes:

1. Digital Asset Management
2. Institutional Repository
3. Digital archives and preservation
4. Content Management System (CMS)
5. Scholarly publishing
6. Digital Library Architecture

FEATURES OF DIGITAL LIBRARIES

Debate rages among librarians and knowledge professionals as to what constitutes a digital library. There are many definitions, ranging from the electronic catalog that describes physical items in a "brick and mortar" library to advanced multimedia environments housing all-digital collections. The Internet has had an incredible impact, but libraries have a history of managing large systems and using technology to deliver bibliographic information.

1. Advanced Search

Digital libraries provide facilities for advanced search to access the information contained in the repository. By advanced search, the extent of desired information can be achieved. The precise searching helps in getting the required results.

* Gain access to the holdings of libraries worldwide through automated catalogs.
* Locate both physical and digitized versions of scholarly articles and books.
* Optimize searches, simultaneously search the Internet, commercial databases, and library collections.
* Save search results and conduct additional processing to narrow or qualify results.
* From search results, click through to access the digitized content or locate additional items of interest.

All of these capabilities are available from the desktop or other Web-enabled device such as a personal digital assistant or cellular telephone. Additionally, the user can customize his or her information request so that search results reflect individual needs and preferences.

2. Browsing of the Digital Library

The user can browse various information products such as electronic journals. The desired information can be taken in the form of print out.

* Journals
* Magazines
* Transactions
* Proceedings
* Newsletters
* Publications by Affiliated Organizations
* Special Interest Groups (SIGs)

3. **Collections of digital scholarly materials** and the services that can help user to find, access, and use them.
4. **Programs that foster innovation** in scholarly communication and digital preservation.
5. **Discovering print and digital materials** in the libraries and beyond, including the million items in the libraries via the digital catalogue.
6. **Streamlining access to online and print materials** by linking from article citations to the full online content or to the locations for printed materials via electronic Links.
7. **Manipulating content** by providing tools to export citations to citation management software, or to create and export presentations using digital images.
8. **Creating and sharing scholarship** by providing faculty with experimental publishing tools and infrastructure.
9. **Influencing commercial products** by developing technical and functional requirements, and by working with subject specialist librarians, known as Resource Liaisons, who monitor and influence licensed products.

In some ways the broadest and most influential digital library efforts have to do with education. Many thousands of teachers, and millions of students, benefit from this resource as it aims to enhance learning through dynamic interaction, visualization, simulation, and other computer-related devices. The Internet and World Wide Web provide the impetus and technological environment for the development and operation of a digital library. The Internet provides the TCP/IP and or its associated protocol for accessing the information and web provide tools and technique for publishing the information over Internet. In the digital environment it is reasonable to say that a central back up or archive should be created at the national level, which will store information out put of the region as well as information from out side the country. Some of the requirement for a digital libraries are given below :

REQUIREMENTS OF THE DEGITAL LIBRARY

To store and retrieve digitized contents, a large amount of storage space and high computing power is required. This has serious implications on the type of hardware required for storage and communication, in terms of the capacity, data rates, bandwidth and type of networks. Also, since search operations are to be performed on images, audio and video, very high throughput computers are required. Preservation and archival objectives require the infrastructure to provide long-term storage solutions.

HARDWARE REQUIREMENTS

* **Computer**: Server, P.C. with multimedia, U.P.S. etc. Computers with enough memory and power to run multiple software programs
* **Server space** to store images and database and also serve on-line databases or web pages
* **Scanner**: Scan jet, flatbed, Sheet feeder, Drum scanner, Slide scanner, Microfilming scanner, Digital camera, Barcode scanner etc. Scanners able to capture scanned images at high resolution and bit depth
* **Storage devices**: Optical storage device, CD-ROM, Jukebox etc. CD/DVD burners, tape back-up systems or extra hard drives to store the high resolution scans for storage and archival purposes
* **Printer**: Laser printer, Dot matrix printer, Barcode printer, Digital graphic printer etc.

SOFTWARE REQUIREMENTS

* Imaging programs like Photoshop to crop, color correct, size and save scans in desired formats
* Database with in-depth fielded information to allow for easy search and retrieval of images
* Web interface to present images and information in fast return

* Fetch, FTP or Appletalk as vehicle for moving images and text Personnel
* Database administrator
* Server administrator
* Scanning and cataloging assistance

PHYSICAL LOCATION

* Secure environment
* Sufficient electrical and networking outlets
* Correct seating and lighting arrangements

STANDARDS

* Standard vocabularies like the Library of Congress Subject Headings should be used for keywords, titles and other descriptive fields
* Image files should be given numeric and unique names (i.e. 001, 002 etc.)
* Image file sizes for the archival scan and all derivatives must be determined
* File formats such as TIFF, JPEG, GIF etc. should be decided upon and then used consistently

  For Example :

  The file sizes and formats used in a Visual Digital Library are:

Archival

* 2025 dpi
* About 2000 x 3000 pixel size
* Between 10 and 18 megabytes in size
* TIFF format
* Stored on CD's

Projection (Optional - Created for in Class Projection)

* 500 dpi
* 700 pixels maximum height or width

* Between 120 kilobytes and 1 megabyte in size
* JPEG format

Screen size (Reference - Meant for Web Viewing)

* 72 dpi
* 500 pixels maximum height or width
* Between 17 and 470 kilobytes in size
* JPEG format

Thumbnail

* 72 dpi
* 100 pixels maximum height or width
* Between 2 and 34 kilobytes in size
* JPEG format
* Available on the web

DIGITAL LIBRARIES AND IMAGES

Converting collections into digital images begins with project planning, and then ostensibly follows a linear progression of tasks; select/prepare, convert/catalog, store/process, distribute and maintain over time. It is risky, however to proceed with any step before fully considering the relationship between the conversion- where quality , throughput and cost are primary consideration and access where process, speed and usability are desirable. Informed project management recognizes the interrelationship among each of the various processes and appreciates that decisions made at the beginning affect all subsequent steps. An excessive concern with user needs, current technological capabilities, and image quality or project costs alone may compromise the ultimate utility of digital collections. At the outset, therefore, those involved in planning a conversion project should ask, "How good do the digital images need to be to meet the full range pf purposes they are intended to serve?" Once the general objectives for quality and functionality have been set, the following factors will collectively determine whether or not these benchmarks will be met:

1. the nature if the source documents themselves, particularly in terms of their individual and collective value
2. the strengths and weaknesses of the current state of digital image technology
3. the institution's technical capabilities and commitment, including staffing
4. current and future users' needs and expectations
5. all relevant budgetary considerations

The best means to ensure an image collection's longevity is to develop a standard of archival quality for scanned images that incorporates the functions range of an institution's services and programs. It should be possible to use the archival image in any of the contexts in which (the source) would be used, for example, for direct viewing, prints, posters and publications as well as to open new electronic opportunities and outlets. It is true that the value of the digital images depends upon their utility, then one must recognize that this is a relative value defined by a given sate of technology and immediate user needs. Digital library design should remain flexible to accommodate the new practices and new document types that are likely to emerge over the coming years.

It is difficult to conceive of the full range of image processing and presentation applications that will be available in the future but if is safe to say they will be considerable. Image formats, compression schemes, network transmission, monitor and printer design, computing capacity and image processing capabilities particularly for the automatic extraction of metadata (OCR) and visual searching (QBIC) are all likely to improve dramatically over the next decade. One can expect each of these developments to influence user needs and lead users to expect more of electronic information. As long as functionality is measured against how well an image performs on "today's technology" then it is expected the value of the digital collections currently created to decrease over time.

Computer vision meets digital libraries when there are lots and lots of images. The interest is two fold.

* Providing access to large image datasets through browsing and search.
* Using large image collections as data for classical computer vision problems such as object recognition.

These two problems are closely related because images have both **semantic** and **visual** content. The approach is to build statistical models which "explain" the data in a collection. The data consists of image segment features and words associated with the images. The words may be carefully chosen keywords as in the Corel data set, or free-form text in conjunction with natural language pre-processing. Sometimes associated words provide semantic content which is difficult to derive using standard computer vision methods. Conversely, the image features provide information which is often omitted when users provide the words because it is clearly a visual element. For example, an image of a red rose will not normally have the keyword "red". Thus image features and associated words can complement and even **disambuguate** each other.

Once a statistical model has been built from the data, it can be queried in a probablistic sense. Specically, one can ask which images have high probability given the query items, which can be any combination of words and image features. An extreme use of such queries is what is referred to as **"auto-illustrate".** Given a text passage, one can query an image data base for appropriate images. Going even further, the model can be queried to attach words to pictures—**"auto-annotate"**, which has clear links to recognition.

The next step towards recognition is to attach the words to specific image regions. To do this it is must to solve a correspondence problem while training because which image parts the words refer too is not known, a priori. The situation is analogous to that in learning how to do machine translation translation from aligned bi-texts. Here user has sentences assumed to have the same content, but the n'th word in one does not necessarily translate to the n'th word in the other. Similarly, the approach of recognition as **translating** from visual cues to semantic correlates (words). The concept of a "digital library" is not merely equivalent to a digitized collection with information management tools. It is rather an environment to bring together collections, services, and people in support of the full life cycle of creation, dissemination, use, and preservation of data, information, and knowledge."

BUILDING OF DIGITAL LIBRARIES

In the age of the Internet and the World Wide Web, the digital libraries have suddenly become hot. Due to expanding of Internet, particularly the WWW, more people are recognizing the need to search indexed collections. The term "digital" is actually somewhat of a misnomer. Digital libraries basically store materials in electronic format and manipulate large collections of those materials effectively. So research into digital libraries is really research into network information systems. The key technological issues are how to search and display desired selections from and across large collections. While practical digital libraries must focus on issues of access costs and digitization technology, digital library research concentrates on how to develop the necessary infrastructure to effectively mass-manipulate the information on the Net.

One of the largest issues in creating digital libraries will be the building of digital collections. There are essentially three methods of building digital collections.

1. Digitization, converting paper and other media in existing collections to digital form.
2. Acquisition of original works created by publishers and scholars.
3. Access to external materials not held in house by providing pointers to web sites, other library collections, or publishers' servers.

DIGITAL LIBRARY MANAGEMENT SYSTEM

The purpose of the information architecture is to represent the riches and variety of library information, using the building blocks of the digital library system. From a computing view, the digital library is built up from simple components, notably **digital objects**. A digital object is a way of structuring information in digital form, some of which may be **metadata**, and includes a unique identifier, called a **handle**. However, the information in the digital library is far from simple. A single work may have many parts, a complex internal structure, and one or more arbitrary relationships to other works. To represent the complexity of information in the digital library, several digital objects may be grouped together. This is called a **set of digital objects**. All digital objects have the same basic form, but the structure of a set of digital objects depends upon the information it represents. The different types of material in a digital library, information can be divided into **categories**, e.g.: text with SGML mark-up, World Wide Web objects, computer programs, or digitized radio programs. Within each category, rules and conventions describe how to organize the information as sets of digital objects. For example, specific rules will describe how to represent a digitized radio program. For each category, the rules describe the digital objects that are used to represent material in the library, how each is represented, how they are grouped as a set of digital objects, the internal structure of each digital object, the associated metadata, and the conventions for naming the digital objects.

User interface that is aware of the rules and conventions applying to certain categories of information is able to interpret the structure of the set of digital objects. Complex information can be presented without the user having any knowledge of the complexity. Since the user interface recognizes how material is represented, it can provide unsophisticated users with flexible access to rich and complicated information.

Digital Library Management Systems (DLMSs) generally

manage collections of multi-media digitalized data and include components that perform the storage, access, retrieval, and analysis of the collections of data. Recently, the new trend of DLMS applications is pushing towards a components/services technology which is becoming more and more standardized. The results of this new orientation are ad-hoc solutions for different components and services of DLMS: the data repository, the data manager, the search and retrieval components, etc.

User interfaces

Both the pilot and the prototype have two user interfaces: one for the users of the library, the other for the librarians and system administrators who manage the collections. Each user interface is in two parts. A standard Internet **browser** is used for the actual interactions with the user. This can be Netscape Navigator, Microsoft's Internet Explorer. The browser connects to **client services**, which provide intermediary functions between the browser and the other parts of the system. The client services allow the user to decide where to search and what to retrieve; they interpret information structured as digital objects; they negotiate terms and conditions, manage relationships between digital objects, it is important to remember the state of the interaction, and convert among the protocols used by the various parts of the system.

Repository

Repositories store and manage digital objects and other information. A large digital library may have many repositories of various types, including modern repositories, legacy databases, and Web servers. The interface to this repository is called the **repository access protocol (RAP)**. Features of RAP are explicit recognition of rights and permissions that need to be satisfied before a client can access a digital object, support for a very general range of disseminations of digital objects, and an open architecture with well defined interfaces.

Handle system

Handles are general purpose identifiers that can be used to identify Internet resources, such as digital objects, over long periods of time and to manage materials stored in any repository or database. When used with the repository, the handle system receives as input a handle for a digital object and returns the identifier of the repository where the object is stored.

Search system

The design of the digital library system assumes that there will be many indexes and catalogs that can be searched to discover information before retrieving it from a repository. These indexes may be independently managed and support a wide range of protocols.

THE STRUCTURE OF INFORMATION IN A DIGITAL LIBRARY

Interactions require that information in a digital library be organized effectively. Within the library, information is stored as basic units of digital information, e.g., a digitized map, a section of text, a Web page, a scanned photograph, etc. In digital form, each basic unit is a sequence of bits, but users often want to refer to material at a higher level of abstraction than the individual item. Common English terms, can refer to many items that are variants of each other. They may have different formats, minor differences of content, different usage restrictions, and so on, but for some purposes users are willing to consider them as equivalent.

The issues to be addressed in structuring information include the following.

* Digital materials are frequently related to other materials by relationships such as part/whole, sequence, etc. For example, a digitized text may consist of pages, chapters, front matter, an index, illustrations, and so on. In the World Wide Web, a typical item may include several pages of text, with embedded images, and links to other

information. A single computer program is assembled from many files, both source and binary, with complex rules of inclusion. Materials belong to collections.

* The same item may be stored in several digital formats. Sometimes, these formats are exactly equivalent and it is possible to convert from one to the other. At other times, the different formats contain different information.
* Since digital objects are easy to change, different versions are created continually. Versions may differ by a single bit or may be very different. When existing material is converted to digital form, the same physical item may be converted several times.
* Each element of digital information may have different **rights and permissions** associated with it.
* The manner in which the user wishes to access material may depend upon the characteristics of **computer systems and networks**, and the size of the material. For example, a user connected to the digital library over a high speed network may have a different pattern of work from the same user when using a dial-up line.

EXAMPLES OF DIGITAL LIBRARIES

1. California Digital Library

The California Digital Library (CDL) supports the assembly and creative use of the world's scholarship and knowledge for the University of California libraries and the communities they serve. In addition, the CDL provides tools that support the construction of online information services for research, teaching, and learning, including services that enable the University libraries to effectively share their materials and provide greater access in digital content.

Collections

The University of California's scholars and students conduct their research and learning while drawing from a significant portion of the world's knowledge held in the library collections. The CDL leads a collaborative effort to license electronic scholarly materials on behalf of the University in co-investment with the campus libraries. In addition, the CDL hosts digital content created by the libraries and other cultural heritage organizations, and makes these materials available for public use.

The CDL:

* **Provides the academic community with more digital material** than is available at any other university, including more than 21,000 electronic journals, tens of thousands of electronic books, and more than 250 article and reference databases.
* **Saves millions of dollars** compared to what would be spent by the campuses acting independently.
* **Allows campus libraries to focus** on local research and curriculum by economizing on costs for commonly needed materials and acquisition processes.
* **Provides a collection of online cultural heritage materials** from California libraries, museums, and historical societies.
* **Enables the experimental publication** of scholarly output in open-access repositories.
* **Encourages interactive use** of government and social science statistics and data.
* **Forges partnerships** with university presses and societies to digitally publish scholarly works.

2. The Alexandria Digital Library

The Alexandria Digital Library (ADL) is a distributed digital library with collections of georeferenced materials. ADL includes the operational library, with various nodes and collections, and the research program through which digital library architectures, gazetteer applications, educational applications, and software components are modeled, prototyped, and evaluated. ADL provides HTML clients to access its collections and gazetteer, and provides specific information management tools, such as the Feature Type Thesaurus for classing types of geographic features, as well as downloadable software code.

3. Bodleian digital Library

This library links to some manuscripts for which digital coverage is quite extensive, providing an introduction to the script, decoration, and musical notation of medieval manuscripts, and the iconography of images illustrating various texts.

List of manuscripts, arranged by century and country of origin:

* 11th century
* 12th century
* 13th century
* Late 13th to early 14th century
* 14th century
* Late 14th to early 15th century
* 15th century
* Late 15th to early 16th century
* 16th century
* 17th century
* Miscellaneous images of 8th- to 19th-century items from the old 'Towards an image catalogue' web pages, now re-scanned

It also provides a collection of images focusing on the 15 boxes of Motor Car ephemera in the John Johnson Collection, supplemented by 1,000 images of other forms of transport. The Internet library of early journals is also developed. An eLib Project was carried out to create a digital library of 18th and 19th Century journals

Bodleian Library Broadside Ballads Project: The Bodleian Library has unparalleled holdings of over 30,000 ballads in several major collections. The original printed materials range from the 16th- to the 20th-Century.

ADVANTAGES OF DIGITAL LIBRARIES

Traditional libraries are limited by storage space; digital libraries have the potential to store much more information, simply because digital information requires very little physical space to contain it. As such, the cost of maintaining a digital library is much lower than that of a traditional library. A traditional library must spend large sums of money paying for staff, book maintenance, rent, and additional books. Digital libraries do away with these fees. Digital libraries can immediately adopt innovations in technology providing users with improvements in electronic and audio book technology as well as presenting new forms of communication such as wikis and blogs.

* **No physical boundary**. The user of a digital library need not to go to the library physically; people from all over the world can gain access to the same information, as long as an Internet connection is available.
* **Multiple accesses**. The same resources can be used at the same time by a number of users.
* **Information retrieval**. The user is able to use any search term bellowing to the word or phrase of the entire collection. Digital libraries can provide very user-friendly interfaces, giving clickable access to its resources.
* **Preservation and conservation**. An exact copy of the original can be made any number of times without any degradation in quality.
* **Space**. Whereas traditional libraries are limited by storage space, digital libraries have the potential to store much more information, simply because digital information requires very little physical space to contain them. When a library has no space for extension digitization is the only solution.
* **Structured approach**. Digital libraries provide access to

much richer content in a more structured manner, i.e. we can easily move from the catalog to the particular book then to a particular chapter and so on.

* **Round the clock availability**. A major advantage of digital libraries is that people can gain access to the information at any time, night or day.

* **Networking**. A particular digital library can provide a link to any other resources of other digital libraries very easily; thus a seamlessly integrated resource sharing can be achieved.

* **Cost**. In theory, the cost of maintaining a digital library is lower than that of a traditional library. A traditional library must spend large sums of money paying for staff, book maintenance, rent, and additional books. Although digital libraries do away with these fees, it has since been found that digital libraries can be no less expensive in their own way to operate. Digital libraries can and do incur large costs for the conversion of print materials into digital format, for the technical skills of staff to maintain them, and for the costs of maintaining online access (i.e servers, bandwidth costs, etc.). Also, the information in a digital library must often be "migrated" every few years to the latest digital media. This process can incur very large costs in hardware and skilled personnel.

Some of the key advantage digital libraries provide are as follows.

(a) Universal accessibility of data.

(b) Data can be access by all type of uses at all level retrieving all kind of information at a time.

(c) Information is provided in a large amount than it possible to physical acquired and maintained as compares to traditional and automated library.

(d) The E-books and journals provide keywords, subject and various other searches.

(e) Facility for downloading and printing.

(f) Saving cost-man power for publishing and brining out new edition.

(g) One copy of document could be viewed by many numbers of persons simultaneously.

Digital libraries offer many advantage as discussed above. However, it is also poses challenges and problems. Digital libraries contain information collection predominantly on digital of electronic form. Electronic publications have some special problems of management as compared to printed documents. This includes infrastructure, acceptability, preservation, copyright, user interface.

DISADVANTAGES OF DIGITAL LIBRARIES

Though many advantages are associated with digital libraries yet some people have criticized that digital libraries are hampered by copyright law, because works cannot be shared over different periods of time in the manner of a traditional library. The content is, in many cases, public domain or self-generated content only. Some digital libraries, such as Project Gutenberg , work to digitize out-of-copyright works and make them freely available to the public. An estimate of the number of distinct books still existent in library catalogues from 2000 B.C. to 1960, has been made.

Digital libraries cannot reproduce the environment of a traditional library. Many people also find reading printed material to be easier than reading material on a computer screen although this depends heavily on presentation as well as personal preferences. Also, due to technological developments, a digital library can see some of its content become out-of-date and its data may become unaccessible. Access to digital libraries and their collections is dependent upon a stable information technology infrastructure (power, computers, communications links etc.).

SOME PROBLEMS ASSOCIATED WITH DIGITAL LIBRARIES

Though digital libraies have opened a new door in the information services by overcoming the day to day problems yet some aspects cannot be ignored. The computer viruses, lack of standardization for digitized information, quick degrading properties of digitized material, different display standard of digital product and its associated problem, health hazard nature of the radiation from monitor etc. makes digital libraries at times handicap.

1. Copyright : Digitization violates the copy right law as the thought content of one author can be freely transfer by other without his acknowledgement. So one difficulty to overcome for digital libraries is the way to distribute information. How does a digital library distribute information at will while protecting the copyright of the author?

2. Speed of access : As more and more computer are connected to the Internet its speed of access reasonably decreasing. If new technology will not evolve to solve the problem then in near future Internet will be full of error messages.

3. Initial cost is high : The infrastructure cost of digital library i.e. the cost of hardware, software; leasing communication circuit is generally very high.

4. Band width : Digital library will need high band for transfer of multimedia resources but the band width is decreasing day by day due to its over utilization.

5. Efficiency : With the much larger volume of digital information, finding the right material for a specific task becomes increasingly difficult.

6. Environment : Digital libraries cannot reproduce the environment of a traditional library. Many people also find reading printed material to be easier than reading material on a computer screen.

7. Preservation : Due to technological developments, a digital library can rapidly become out-of-date and its data may become inaccessible.

Major administrative challenge is in complying with copyright and intellectual property right issues. The library authority has to discuss seriously with publishers on this aspect to have profit to both uses, publishers as well as authors users may be charged for each access, downloading from servers and each kind of digital library collection. This would provide for publishers, authors and libraries. Security aspects is one of the most pressing challenge of digital affairs piracy old data-base viral invasions, parallel satellite networking stress are some of the issues for digital libraries are confronted as a way of routine. Due to lack of expertise in the country and abroad overseas vendors change too much and also reluctant to import techniques/ technology.

ESSENTIAL ASPECTS TO DELIVER SERVICES

Though digital libraies seem to overcome the cumbersome acitvities of traditional libraies yet many approches are required to make it effective and fully utilized. Some of the important actions are as following:

1. Curbing the costs : The continuing substantial costs of providing electronic information will require that electronic collection decisions be made even as carefully and parsimoniously as for print.

2. Backup mechanisms : Backup/restore procedures must be in place and must be automated and economical, for libraries are never likely to have expensive labor available in quantity. Backups must be multi-generational, using remote storage, with regular disaster simulations and tests.

3. Staged Access : In computing jargon, "staging" refers to the prioritized use of different mechanical methods of storing data as it waits to be recalled. All data does not need to be immediately available on the most expensive and fastest storage media. Alternatives for providing immediate online access to the enormous potential volume of scholarly information need to be provided. What can be off line, and how can it be retrieved? Present alternatives include magnetic disks, optical disks and jukeboxes, optical disks on shelves, magnetic tapes on site, tapes in remote storage, and automated data warehouses of magnetic tapes.

4. Data structure standards : In a repository, does information simply exist as is (as first created) or is complementary information (metadata) associated with it? Widely differing examples include SGML (Standard Generalized Markup Language) headers, ICPSR codebooks, picture captions, hypertext links and early software versions for use with data files. There is an increasing need to link bit-mapped page images to ASCII text versions of the page contents.

5. Refreshing mechanisms : Refreshing is agreed to be necessary for long-term preservation across advances in computing technology, media and software. There will be organizational and bureaucratic issues in addition to the simply technical. If information is copied from magnetic to optical disk, copyright issues must be recognized. Automation will be necessary to reduce labor costs. Other issues include workflow and record-keeping, migration techniques, and standards and techniques that will apply independently of technology. It may be possible to link refreshment to backup techniques for expedience and economy.

6. Authentication and Integrity : Intellectual preservation goes beyond preservation of the medium and the technology to assure the protection of the intellectual structure of information as it was recorded by its author. For example, bit patterns of texts, sound and images may be preserved through cryptographic hashing and encoding methods such as the digital time-stamping technique. Standards and conventions for use and citation will be necessary.

7. Redundancy : It will be important to establish

standards for the number of repository locations necessary to assure long-term existence of specific electronic information and access to it. One location won't do for a particular major electronic document or set; will two, or three? How many? Major institutions may separately or consortially establish repositories. It is not yet clear how much redundancy of their components will be desirable among them.

Extension of Digital Library Services : Extension of digital library services includes:

1. Helping the other university libraries put digital collections online
   * Publication of Archival, Library & Museum Materials
   * Support for Electronic Theses and Dissertations (ETDs)
   * Standards and guidelines for digitization, metadata, and mark-up.
   * Cost-effective central storage facilities
   * Tools for loading, storing, searching, and displaying digital objects
2. Grant writing and coordinating grant funded projects
   * Opening Archives: Improving Access to Primary Sources Enhanced Access to Special Collections
3. Hosting digital collections and indexes
4. Providing tools for managing digital resources over time
   * Digital Archive
   * Usage statistics for digital collections
   * Naming services — server
   * Open Archives Initiative (OAI) data provider (broker)
   * Open Archives Initiative (OAI) service provider
5. Making scholarly journals accessible on the web

REFERENCES

1. Ansari, Mehtab Alam (2003) Digital libraries: needs, technology and benefit, ILA Bulletin , 38(3), Pp 22-26.
2. Banerjee, Swapna and Chakrabarty, Biplab (1999), Digital libraries: some issues and perspective, ILA Bulletin 34(3-4), October 1998 to March 1999, Pp. 60-63.
3. Cyric, Jiji, Deshmukh, G.R. and Rajalakshmi(2002) , Digitizationof libraries: in modern era, ILA Bulletin 38(3) Pp.68-73.
4. portal.acm.org/dl.cfm
5. Dhaka, R.P.S. and Arora, Kamlesh (1995), Electronic libraries: A myth or a reality, Annals of library science and documentation.42(4). Pp.152-59.
6. *www.digitallibrary.ernet.in*
7. www.nalanda.nitc.ac.in
8. Khan, Shakeel Ahmad, Khayat, Roshan and Yunus, Mohad (2003), Digital libraries: the present scenario, ILA Bulletin 39(2) Pp.3-7.
9. dli.iiit.ac.in
10. digital.library.upenn.edu
11. *www.dlib.org/*
12. Bates, M.J. (1988) "How to Use Controlled Vocabularies more Effectively in on-line Searching" *On-line* 12(6), November, 45
13. Bates, M.J. (1998) "Indexing and access for digital libraries and the Internet: human, database and domain factors". *Journal of the American Society for Information Science* 49, 1185-1205.
14. Chang, S., Smith, J.R., Meng, H.J., Wang, H. and Zhong, D. (1997) "*Finding Images/Video in Large Archives: Columbia's Content-Based Visual Query Project*". *D-Lib Magazine*. February. Available at: http://www.dlib.org/dlib/february97/columbia/02chang.html
15. *www.ietdl.org*
16. *www.academicinfo.net/digital.html*
17. www.ifla.org/II/metadata.htm
18. Kumar, Bhuvan and Reddy, V.Sreeniuas (2003), Digital library use: a case study of NIT library, Waraangal, ILA Bulletin 39(2) Pp.40-45.
19. *www.muktabodha.org/digital_library.htm*
20. paniit.iitd.ac.in/indest
21. www.nls.uk/digitallibrary
22. Ravichandra Rao I.K. and Suma P. (1996), Digital libraries, challenge

and issues, Digitized information paper presented at the SIS-96, 18-20 January 1996, Bangalore, (Eds) T.B.Rajasekhar, I.K. Ravichandra Rao and N.V. Satyanarayana, Pp.185-95.

23. www.diglib.org
24. *http://en.wikipedia.org/wiki/Digital_library*
25. *http://scholar.lib.vt.edu/DLI2/defineDL.html*
26. Poulter, A. (1997). "The design of World Wide Web search engines: a critical review". *Program* 31(2), 131-145.
27. Quint, B. (1991, July) "Inside a Searcher's Mind: The Seven Stages of an on-line Search - Part 2" *On-line*. 15(4), 28-36.
28. Schwartz, C. (1993) "Evaluating CD-ROM products: yet another checklist". *CD-ROM professional* 6(1), 87-91.
29. Schwartz, C. (1998) "Web search engines". *Journal of the American Society for Information Science*. 49(11), 973-982.
30. *http://www.clir.org/pubs/issues/issues04.html#dlf*
31. *http://www.dlib.org/metrics/public/papers/dig-lib-scope.html*
32. *http://www.firstmonday.org/issues/issue6_6/lynch/#l2*
33. *http://www.kmentor.com/socio-tech-info/archives/000694.html*
34. R. Pool, "Turning an Info-Glut into a Library," *Science*, Oct. 7, 1994, pp. 20-22.
35. G. Taubes, "Indexing the Internet," *Science*, Sept. 8, 1995, pp. 1,354-1,356.
36. Chandran, D. Digital Libraries: Guidelines for Planning and Implementation. SIS Proceedings, 2003, pp. 40-46.
37. *http://namami.nic.in/digitization.htm*
38. Garg, Babita. Digital Library. SIS Proceedings, 2003, pp. 47-56.
39. *http://www.ifla.org/II/diglib.htm*
40. en.wikipedia.org/wiki/Digital_library
41. Misra, Jyoti. Digital Libraries or a Dilemma. SIS Proceedings, 2003, pp. 57-62.
42. *www.dlib.org/*
43. paniit.iitd.ac.in/indest/workshop/workshop_ondigitallibraries_iitr.html
44. *www.iimahd.ernet.in/~jajoo/dli/dli.htm*
45. Verma, R. Digitization: A Requisite for Digital Library. SIS Proceedings, 2003, pp. 93-100.
46. openmed.nic.in/12

47. Sharma, M. Digital Libraries- An Insight into the Future. SIS Proceedings, 2003, pp. 75-79.

48. www.dlib.org/dlib/february99/02journalreview.html

49. Pandey, K M; Sinha, V K and Kumar, D. Recent Advances in Library Technology, SIS Proceedings, 2003, pp. 511-524.

50. Geoffrey, Nunberg. The Digital Library: A Xerox PARC Perspective, Source Book on Digital Libraries.

8

Physical Aspects of Digital Libraries

DIGITAL LIBRARY TECHNOLOGY TRENDS

Digital library technology trends reflect that the components might not all be part of a discrete digital library system, but could be provided by other related or multi-purpose systems or environments. Accordingly, integration is a consistent issue cited by digital library developers. It is found that interoperating with the existing library infrastructure, the digital library must be designed to work with existing library catalogs and incorporate industry standards, formats, and protocols. The term "digital library" is often used to describe any multimedia management system holding digitized information, but this does not mean it will deliver true library application functionality. Thus, these digital library components must also be tailored to capture, encode, and deliver information according to the standard practices adopted by the library industry. Due to the rapid pace of technological change, some standards are concrete and others are emerging.

A fully developed digital library environment involves the following elements:

1. Initial conversion of content from physical to digital form.

2. The extraction or creation of metadata or indexing information describing the content to facilitate searching and discovery, as well as administrative and structural metadata to assist in object viewing, management, and preservation.
3. Storage of digital content and metadata in an appropriate multimedia repository. The repository will include rights management capabilities to enforce intellectual property rights, if required.
4. Client services for the browser, including repository querying and workflow.
5. Content delivery via file transfer or streaming media.
6. Patron access through a browser or dedicated client.
7. A private or public network.

FUNCTIONAL COMPONENTS OF DIGITAL LIBRARIES

Varieties of digital libraries are being developed today yet most of them share a few common functional components. A basic understanding of the key functional components will help to prepare better in digital library development efforts.

Selection and acquisition : Typical processes covered in this component include: selection of documents to be added, digitization and/ or conversion of these documents to appropriate digital form.

Organization : Key process involved in this component include assignment of metadata (e.g. bibliographic information) to each document being added to the collection.

Indexing and storage : This component carries out indexing and storage of documents and metadata, for efficient search and retrieval.

Repository : This is the core component of the digital library, consisting of the document objects, metadata and indexes created for purposes of search and retrieval.

Search and retrieval : This is the digital library front-end used by the end-users to browse, search, retrieve and view the contents of the digital library. This is typically presented to the users as an HTML page.

Digital library website : This is the server computer that hosts the digital library collection, and presents the collection to the user in the form of a website home page. The user selects a suitable link on this page to go to the search and retrieval front-end mentioned above. The digital library delivers the content based on search and retrieval operations. The digital library home page itself may be integrated with the library website through an appropriate hypertext link.

Network connectivity : For online access, the digital library website computer should have dedicated connection to the intranet and/or Internet. Depending on the target user community, access may be restricted to the intranet (organizational LAN) or extended to external users through the Internet.

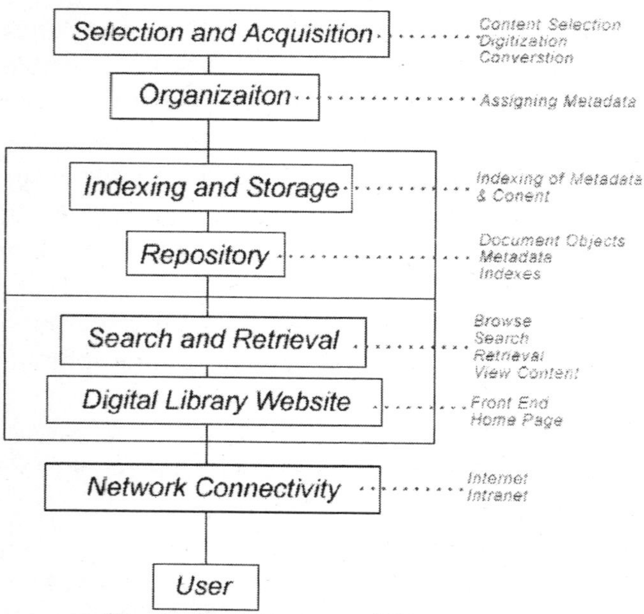

(Source: http://www.ncsi.iisc.ernet.in/)

ARCHITECTURE OF DIGITAL LIBRARY

Digital objects

The actual content of digital libraries is made up of a number of digital objects. In some cases these may be thought of as data sets (e.g., a table of results, the genomic information for an individual). In others they may be multimedia information, such as an image, graphic, animation, sound, musical performance, or video. Many can be thought of as documents, which carry content in some structure or structures, perhaps made up of logical or physical divisions such as sections or pages. Some of the objects will be "born digital", while others may be representations of some physical object that result from some type of digitization process. Thus, into the foreseeable future, digital libraries will be hybrid constructs, where paper, microforms, and other media carry much of the content that is of interest and only the metadata is in digital form.

Metadata

Digital objects are described, structured, summarized, managed, and otherwise manipulated in surrogate form through the use of "metadata", which literally means data about data. Three types of metadata are often distinguished: descriptive, structural, and administrative. Metadata is usually produced through a process called "cataloging" that is often carried out by trained librarians. Collections of such information are commonly stored in "catalogs". In computerized environments, metadata may be automatically or semi-automatically extracted or derived from the original content, or the "full-text" may simply be indexed and searched without involving metadata. Nevertheless, if metadata is available and can be used along with content terms derived from full-text documents, the result are better than if only one source of evidence is employed. Yet, if only metadata is available in computer form to describe a digital object, it must be used in digital libraries. Hence, metadata should be used whenever available in a digital library, and is an important aspect in many such systems.

Repositories and Harvesting

Digital libraries are containing a collection of digital objects (DOs), each of which has one or more sets of metadata objects (MDOs) associated. This "repository" part of a digital library may, as is the case in the Open Archives Initiative, follow certain conventions. In particular, according to the latest specifications, an "Open Archive" (OA) is a computer system with a WWW server that behaves according to an OA protocol to allow other computers to harvest metadata from it. That protocol supports requests to, for example:

* list what types of metadata format are present,
* list what structure of sets and subsets are used to organize or partition the content,
* disseminate or return a particular MDO, or
* list URIs (unique identifiers) for all MDOs added during a particular date range.

INDEXING, RESOURCE DISCOVERY, SEARCHING, AND RETRIEVING

Indexing may build upon any manual indexing carried out by authors, other creators, or indexers. Automatic indexing also may involve first classifying DOs, such as when OCLC's CORC project software suggests Dewey Decimal Classification entries for a WWW page that is being cataloged. Indexes may be centralized or distributed. They may be two-level, allowing a resource discovery phase to proceed to find what source(s) should be included in the subsequent (lower) level search. Indexes also may have multiple parts, such as when a document has text, image, audio, or video content. Content-based indexing of multimedia information generally involves identifying and assessing features that characterize the DOs, whether they involve concepts, n-grams, words, keywords, descriptors, phonemes, textures, color histograms, eigenvalues, links, or user ratings. Most commonly, searching in a digital library involves an information retrieval system or search engine. In some cases a database management system is used instead or underlies the retrieval system. In any case, retrieval will be more effective if a suitable scheme is used to combine the various types of evidence available, to indicate if a DO may be relevant with respect to the query that is used to express the user's information need.

LINKING, ANNOTATING, AND BROWSING

Once a DO is found, it is appropriate to follow links from it to cited works (or vice versa). Further, notes can be recorded as annotations and linked back to the works, so they can be recalled later or shared with colleagues as part of collaborative activities. If suitable clustering is in place, other DOs that are "near" a given work may be examined. Or, using a classification system appropriate for the content domain, users may browse around in "concept space" and link at any point between concepts and related DOs. Browsing also can proceed based on any of the elements in the MDO. Thus, dates, locations, publishers, contributing artists, language, and other aspects may be considered to explore the collection or refine a search.

DIGITIZATION

Cornell University Library defines digital images as: "*electronic snapshots taken of a scene or scanned from documents, such as photographs, manuscripts, printed texts, and artwork. The digital image is sampled and mapped as a grid of dots or picture elements (pixels). Each pixel is assigned a tonal value (black, white, shades of gray or color), which is represented in binary code (zeros and ones). The binary digits ("bits") for each pixel are stored in a sequence by a computer and often reduced to a mathematical representation (compressed). The bits are then interpreted and read by the computer to produce an analog version for display or printing.*

In summary, digitization converts materials from formats that can be read by people (analog) to a format that can be read only by machines (digital). Flatbed scanning, digital cameras, planetary cameras, and a number of other devices can be used to digitize cultural heritage materials.

REASONS FOR DIGITIZATION

The main reasons to digitize are to enhance access and improve preservation. By digitizing their collections, cultural heritage institutions can make information accessible that was previously only available to a select group of researchers. Digital projects allow users to search collections rapidly and comprehensively from anywhere at any time. Digitization can even make the invisible visible. The Nebraska State Historical Society has scanned glass plate negatives and enhanced them in order to see inside the formerly dark doors of pioneers' sod houses.

Digitization can also help preserve precious materials. Making high-quality digital images available electronically can reduce wear and tear on fragile items. This does not mean, however, that digital copies should be seen as a replacement for the original piece. Digital files are not permanent and should be maintained and periodically transferred to new formats. Even after digitization, original documents and artifacts must still be cared for. Preservation remains a secondary benefit of digital projects. If preserving a collection is deemed a higher priority than increasing access to it, a better use of resources would to purchase acid free folders, encapsulate fragile documents, or otherwise improve storage conditions

BENEFITS OF DIGITAL ACCESS FOR COLLECTIONS

1. Can be viewed from anywhere, at any time of the day
2. Can be printed from the web
3. Viewers can find what they are looking for quickly and independently.
4. Can save staff reference time by answering frequently asked questions on the web.
5. Can enhance images electronically so that they can be viewed with greater legibility.
6. Increases use of collections and facilitates learning and scholarship.

The Preservation Benefits for Collections

* Objects do not have to be reshelved or located by staff.
* Objects are not handled frequently which reduces wear and tear.

The Downside of Digitization

Despite everything that digitization can accomplish, there are also some very good reasons to stay out of the digital realm. First, not every collection is worth digitizing. The idea of entire libraries or museums being completely online is a long way off, and many experts say that it will never happen. Successful digital projects are the result of careful evaluation of collections, and the digitization of only those items that will provide the greatest benefit to the user

Digital projects are also expensive. At this point no institution has managed to make digitization projects cost-effective, and most attempts to recoup the costs of digital imaging through user fees have failed. Costs for digitization continue even after a project's conclusion, as all digital files require maintenance to ensure that they will readable in the future.

Though the high costs of digital conversion exist, it is a near certainty that digitization's importance will increase exponentially in the future. Users are already requesting high quality digital images from cultural heritage institutions, and this demand will only increase as time goes on.

Planning for Digitization

The success of digital projects depends not on expensive technology, but rather on sound planning. Perhaps because digitization is relatively new, libraries too often concentrate on technological aspects before deciding on a project's goals. Technology should never drive digital projects. Goals should be determined first, and only then should the appropriate technology be selected in order to meet a project's objectives. Digital projects require new skills. Digitization planning should allow time to teach current staff new technologies. Even if an outside vendor completes a project, or new staff is hired specifically to work on a digital project, permanent staff should at least learn the basic theories and practices of digitization. Libraries often hire short-term staff for digitization projects which can result in the loss of digital expertise when the project ends

GOALS OF THE DIGITAL LIBRARY SOFTWARE

* Identifiers
* persistent unique names for all resources without respect to machine address
* Relationships
* support inter-relationships among objects
* Tame Content
* Normalize heterogeneous content and metadata using extensible object model
* Integrated management
* Manage the digital resources and metadata, as well as supporting computer programs, services and tools
* Interoperable access
* Standard protocol for getting information about objects, and accessing object content
* Discover and execute extensible service operations for objects
* Scalability
* Support for >10 million objects
* Security
* Flexible authentication and policy enforcement
* Preservation
* Features to support longevity and archiving
* XML serialization of objects
* Content versioning
* Content recon

* Re-use: enable content to be present in any number of contexts
* Re-purposing: mechanism for dynamically transforming content to fit new requirements
* Self-actualizing objects
* Allow digital objects to disseminate appropriate launch-pads or tools for end user interaction with content

Features of a Digital Library Software

* Digital Object Model
* **Abstraction :** model for heterogeneous digital resources
* **Flexible :** create different "content models"
* **Generic :** no bifurcation of metadata and content
* **Aggregation :** both locally stored content items and by-reference content items
* **Extensible :** associate services with objects (extensible behavior interfaces)

Distributed Repositories

* Interoperable object serialization (defined by XML schema)
* Common web service APIs for access and management
* Federation for distributed, virtual collections
* Object Lifecycle Management (archive and preservation)
* XML ingest, storage, and export
* Content versioning
* Object-to-Object relationships
* Event history

Content repurposing

* Provide multiple views of content/metadata
* Dynamic transformations of content/metadata
* Add new views/transformations over time

Web Services

* Software is exposed via web services

* Software can interact with other web services
* Software uses WSDL to define interfaces and XML transmission
* Easy integration with other applications and systems
* Does not assume any particular workflow or end-user application
* Generic repository service as substrate for many kinds of applications

Digital Content

* Some conventional objects
* Print material such as books, maps, atlas etc.
* Complex, compound, dynamic objects
* Electronic material, graphic files etc.

Key questions while implementation of digital library

* How can clients interact with heterogeneous collections of complex objects in a simple and interoperable manner?
* How can complex objects be designed to be both generic and genre-specific at the same time?
* How can we associate services and tools with objects to provide different presentations or transformations of the object content?
* How can we associate specialized, fine-grained access control policies with specific objects, or with groups of objects?
* How can we facilitate the long-term management and preservation of objects?

REFERENCES

1. Gambles, Anne. (2000) "Put Yourself in the PIE - The HeadLine Personal Information Environment." *D-Lib magazine,* 6,(4). Available at http://www.dlib.org/dlib/april00/04inbrief.html#GAMBLES
2. Borgman, C.L. (1999) *in: Source Book on Digital Libraries: Report for the National Science Foundation (TR-93-35-(439))* edited by Fox, E A. Blacksburg (VA): VPI and SU Computer Science Department. Available at: http://fox.cs.vt.edu/DLSB.html.

3. Cohen, Suzane, Fereira, John, Horne, Angela, Kibbee, Bob, Mistlebauer, Holly and Smith Adam. (2000) "MyLibrary: personalised electronic services in the Cornell University Library." *D-Lib Magazine,* 6(4). Available at: http://www.dlib.org/dlib/april00/mistlebauer/04mistlebauer.html
4. eLib: Electronic Libraries Programme. (2001) Available at: http://www.ukoln.ac.uk/services/elib/
5. EveryWare Development Inc. (1999) "Tango Enterprises 3: User's Guide." Austin, TX: Pervasive Software. [Note: EveryWare Development Inc. was acquired by Pervasive Software in 1998]
6. Fischer, G. and Lemke, A. (1998) "Construction Kits and Design Environments: Steps Toward Human Problem-domain Communication." *Human Computer Interaction,*3,(3), 179-222.
7. Fox, E.A. (1999) "The Digital Libraries Initiative: Update and Discussion." *Bulletin of the American Society for Information Science,* 7-11.
8. Fox, E.A., Hix, D., Nowell, L.T., Brunei, D.J., Wake, W.C., Heath, L.S. and Rao, D. (1993) "Users, User Interfaces, and Objects: Envison, a Digital Library." *Journal of the American Society for Information Science,*44(8), 480-91.
9. Gould, J. and Lewis, C.(1985) "Designing for Usability: Key Principles and What Designers Think." *Communications of the ACM,* 28, (3), 300-311.
10. iGEMS. (2001). Available at: http://gemsweb.ntu.edu.sg/iGems/public.htm [Accessed 7 August 2001].
11. Networked Digital Library of Theses and Dissertations. (2001). Available at: http://www.theses.org.
12. Oppenheim, C. (1999) "What is the Hybrid Library?" *Journal of Information Science,* 25,(2), 97-112.
13. Wilson, S. and Johnson, P. (1996) "Bridging the Generation Gap: From Work Tasks to User Interface Design." *in: Computer-Aided Design of User Interfaces,* edited by J. Vanderdonckt. Namur: Presses Universitaires de Manur.
14. www.sun.com/products-n-solutions/edu/whitepapers/pdf/digital_library_toolkit.pdf
15. www.thdl.org/tools
16. sunsite.berkeley.edu/Tools
17. www.dlib.org/dlib/january03/kalinichenko/01kalinichenko.html
18. rkcsi.indiana.edu/archive/CSI/WP/wp97-05B.html
19. www.capurro.de/diglib.htm
20. ei.cs.vt.edu/~dlib/def.htm

21. Lesk, M. (1991) "The CORE Electronic Chemistry Library." *14th Annual International ACM/SIGIR Conference on R&D in Information Retrieval,* Chicago, IL. pp. 92-112.
22. Lewis, C. and Rieman J. (1993) "Task-centered User Interface Design - A Practical Introduction." *University of Colorado:* Boulder.
23. Meyyappan, N., Chowdhury, G. and Foo, S. (2000) "A Review of the Status of 20 Working Digital Libraries." *Journal of Information Science,*26, (5), 337-55.
24. Michelle, Q. and Wang, B.(2000) "A User-Centered Interface for Information Exploration in a Heterogeneous Digital Library." *Journal of the American Society for Information Science,* 51,(3), 297-310.
25. Networked Computer Science Technical Reference Library. (2001). Available at: http://www.ncstrl.org.
26. www.computer.org/.../2006/v8n4&file=edu.xml
27. www.csdl.tamu.edu/DL95/papers/nuernberg/nuernberg.html
28. dspace.inflibnet.ac.in/bitstream/1944/443/1/04Planner_31.pdf
29. www.cs.mdx.ac.uk/ridl
30. http://www.library.cornell.edu/preservation/tutorial/intro/intro-01.html
31. jodi.tamu.edu/Articles/v04/i02/Heath
32. www.wtec.org/loyola/digilibs/04_03.htm
33. www.fedora.info/documents/ecdl2002final.pdf
34. www.cnri.reston.va.us/cstr/arch.html
35. www.dlib.org/dlib/february97/cnri/02arms1.html
36. Cousins, S. (1996) "A task-oriented interface to a digital library." *Proceedings of the CHI '96 Conference Companion on Human Factors in Computing Systems: Common Ground* pp. 103-104. Available at: http://www-diglib.stanford.edu/diglib/WP/PUBLIC/DOC44.html
37. DIGILIB. (2001). Available at: http://www.architect.uq.edu.au/digilib/index.html
38. Digital Library Federation (2000). Available at: http://www.clir.org/diglib/dlfhomepage.htm
39. Digital Libraries Initiative. (2001). Available at: http://www.dli2.nsf.gov/dlione/
40. www9.org/final-posters/65/poster65.html
41. alexandria.sdc.ucsb.edu/research/dla/index.htm
42. www.diglib.org/architectures.htm
43. www-dbs.ethz.ch/DELOS
44. ieeexplore.ieee.org/iel4/5492/14771/00670378.pdf
45. http://www.clir.org/pubs/reports/pub80-smith/pub80.html#preservation

46. www.jcdl2004.org
47. www.ietdl.org/IET-CDT
48. www.actapress.com/PaperInfo.aspx
49. http://www.hti.umich.edu/m/moa.new/
50. http://memory.loc.gov/ammem/edhtml/edsndhm.html
51. http://www.nebraskahistory.org/lib-arch/research/photos/digital/index.htm
52. drtc.isibang.ac.in/pipermail/dlrg/2006-November/000854.html
53. www.dlib.org/dlib/december97/ibm/12lotspiech.html
54. dlist.sir.arizona.edu/737
55. www.diglib.org/standards.htm
56. cimic3.rutgers.edu/ieee_dltf.html
57. www.ifla.org/II/metadata.htm
58. www.acm.org/pubs/contents/proceedings/series/siggraph
59. webdoc.gwdg.de/ebook/aw/1999/sun/noerrfinal.pdf
60. www.ieee-tcdl.org/Bulletin/v2n1/raghavan/raghavan.html
61. www.cdacnoida.in/digitlib.asp
62. www.kbs.uni-hannover.de/Arbeiten/Publikationen/2004/brase_TIB_hannover.pdf
63. https:/.../bitstream/1849/407/1/Changing%20Role%20of% 20 Librarians%20in%20Digital%20Library%20Era.doc

9
Internet : Basics

The Internet is a meta-network, a constantly changing collection of thousands of individual networks inter-communicating with a common protocol. The architecture of Internet is described in its name, a short from of the compound word "inter-networking". The architecture is based on specification of the standard TCP/IP protocol, designed to connect any two networks which may be very different in internal hardware, software, and technical design. Once two networks are interconnected, communication with TCP/IP is enabled end-to-end, so that any node on the Internet becomes able to communicate with another node.

For example, an office, a desktop computer might be connected to a local area network with a company connection to a corporate Intranet connected to several national Internet service providers. In general, small local Internet service providers connect to medium-sized regional networks which connect to large national networks, which then connect to very large bandwidth networks on the Internet backbone. The organizations running the Internet backbone operate very high bandwidth networks relied on by governments, corporations, large organizations, and other Internet service providers. Their technical infrastructure often includes global connections through underwater cables and satellite links to enable communication between countries and continents. Each communication packet goes up the hierarchy routing of Internet networks as far as necessary to get to its destination network where local routing takes over to deliver it to the addressee. In the same way, each level in the hierarchy pays the next level for the bandwidth they use, and then the large backbone companies settle up with each other.

USING THE INTERNET

The Internet is a complex space with a rich set of useful features and functions. Knowing how to get the most out of the Internet can help user as much as knowing how to read, maybe more. The following table reflects the time of Internet related technologies.

Technology	*Invented*
Using The Internet	1969
Email	1971
Mailing Lists	1972
MUD's	1979
The Usenet	1980
IRC	1988
The Web	1990

For example, Internet Relay Chat (IRC) enables people all over the world to talk together over the Internet in real-time sessions in virtual rooms. Multi-User Dungeons (MUD's) are one of the most unique and entertaining inventions on the Internet. Each MUD is a virtual universe run by a complex server program which manages a local world in response to the words and actions of human and automated visitors.

KEY INTERNET FEATURES

1. Internet Distribution

The geographic distribution of the Internet is continuing to spread, around the world and even beyond. All of the Internet's technologies — web, newsgroups, email, mailing lists, IRC, MUD's — enable geographically distributed groups of people to communicate. The basic architecture of the Internet is open and fundamentally designed to connect new networks, so it has spread rapidly to interconnect our world and turned it into a true multi-way electronic global village. The rapid geographic distribution of the Internet is making the world just a bit smaller. The Internet is the latest and most powerful such invention, with a current distribution to every corner of this planet, and already inevitably moving into space.

2. Internet Architecture

The Internet is the most robust communications network ever designed as it is able to adapt itself almost instantaneously to damage or outages to individual sections. The Internet has no irreplaceable central control, administration, or authority. The loss of individual computers and networks does not affect its overall reliability.

3. Internet at near Light Speed

The Internet operates at near light speed, which on a planet the size of Earth often practically measures to near real-time. Digital information such as Internet packets travel at 2/3 of the speed of light on copper wire and on fiber optic cables. With fixed near-optimal transmission speed, there are two methods to make Internet networks faster, increase the number of bits that are traveling at once down the connection, or increase the speed at which user switches them from one connection to another at the junction points. Internet routers are getting faster with switching speeds nearing instantaneous, while fiber optics and wireless technologies are enabling networks to send much larger numbers of bits at once.

4. Universal Access

In digital communications media, the vast majority of participants are active creators of information as well as recipients. This kind of symmetry has previously only been found in media like the telephone but while the telephone is almost entirely a medium for private one-to-one communication, computer network applications such as electronic mailing lists, conferences, and bulletin boards, serve as a medium of group or 'many-to-many' communication. The Internet provides *universal access*, giving the same powerful capabilities to everyone who has access to the network. The Internet gives equally to everyone the ability to make information like text, audio, and video accessible to a world wide audience at an extremely low cost, since web site storage space and lots of bandwidth can be rented from web hosting providers. The Internet is universally empowering - everyone can participate.

5. Internet Growth Rates

Internet is growing at a rapid rate since its conception, on a curve geometric and sometimes exponential. Internet is growing exponentially in three different directions — size, processing power, and software sophistication — making it the fastest growing technology humankind has ever developed:

Size : Due to its wide and fast expansions, it is inevitable that half the world's population will have access to the Internet.

Power : Usually computer processors and storage continue to double in power and capacity about every 18 months and providing steadily more powerful computers for use by increasing sophisticated software.

Functionality : Software applications from routing programs to browser applications continually build on previous technology to become more sophisticated with every release, continuously evolving to incorporate new features and capabilities.

6. The Digital Advantage

Digital communications have the *D4 advantage*: "Digital data doesn't degrade". Analog systems and digital systems are like mirror images of each other. Analog systems are usually controlled by physical mechanisms that can be in an infinite number of continuous positions. In contrast to approximate, analog systems, the Internet is a digital medium based on data made up of discrete 1's and 0's. A bit of computer data is not infinitely adjustable, and only has one of two unambiguous states — it is either a 1, or a 0. This limitation has a very important compensating advantage: there is no "drift" that can introduce error. The Internet, like all computer systems, is based on digital data, so that information never changes or becomes distorted over time or in transmission between sites. This is the key feature that makes it possible to construct the very complex software systems that run the Internet, so that a web site doesn't age and become fuzzy or garbled over time, and the characters in an email don't get transposed or mixed up when they are sent over long distances. One of the most important strengths of the Internet is that it's based on one of the simplest concepts — digital 1's and 0's.

7. Internet Security

For Internet security, following measures are taken:

Firewall : Firewall is a system or group of systems that enforces an access control policy between two or more networks. The actual means by which this is accomplished varies widely, but in principle, the firewall can be thought of as a pair of mechanisms: one which exists to block traffic, and the other which exists to permit traffic. Some firewalls place a greater emphasis on blocking traffic, while others emphasize permitting traffic. Probably the most important thing to recognize about a firewall is that it implements an access control policy. Many traditional-style corporations and data centers have computing security policies and practices that must be followed. In a case where a company's policies dictate how data must be protected, a firewall is very important, since it is the embodiment of the corporate policy. Firewall provides not only real security—it often plays an important role as a security blanket for management. Some firewalls permit only email traffic through them, thereby protecting the network against any attacks other than attacks against the email service. Other firewalls provide less strict protections, and block services that are known to be problems. Generally, firewalls are configured to protect against unauthenticated interactive logins from the "outside" world. This, more than anything, helps prevent vandals from logging into machines on the network. More elaborate firewalls block traffic from the outside to the inside, but permit users on the inside to communicate freely with the outside.

Virus Protection : Computer viruses are everywhere. Their very names give computer users a scare — Klez, Melissa, W32, Michelangelo. With the use of the Internet and endless communication between computers, viruses are spreading faster than ever. The only way to prevent their growth is through public awareness of safe computing. computer virus is simply a set of computer instructions or computer code that is written by some unscrupulous person. This code is attached to some part of the normal computer operating system or computer program. Instructions in this code tell the computer to perform some task. This task is often a destructive one, such as deleting important information or crashing the hard disk. However, there are other benign viruses that may be only a slight aggravation. Some viruses have been known to do nothing more than put a large happy face on the computer screen. In order to take advantage of these new antivirus shields, one must update the antivirus program on a regular basis. The company that manufactured the antivirus software will provide constant updates as they find new viruses. One should make arrangements to get these updates regularly to be sure that the computer is completely protected. One can go to the Web site of the antivirus manufacturer to download new virus updates, or definitions as they are often called. Some newer antivirus programs automatically update the antivirus protection as well as the antivirus definitions when user logs onto the Internet through a process called live update or online virus scan.

Encryption : A secure computing environment would not be complete without consideration of encryption technology. The term encryption refers to the practice of obscuring the meaning of a piece of information by encoding it in such a way that it can only be decoded, read and understood by people for whom the information is intended. It is the process of encoding data to prevent unauthorized parties from viewing or modifying it. A secure computing environment would not be complete without consideration of encryption technology. The term encryption refers to the practice of obscuring the meaning of a piece of information by encoding it in such a way that it can only be decoded, read and understood by people for whom the information is intended. It is the process of encoding data to prevent unauthorized parties from viewing or modifying it. *Encryption* is said to occur when data is passed through a series of mathematical operations that generate an alternate form of that data; the sequence of these operations is called an *algorithm*. To help distinguish between the two forms of data, the unencrypted data is referred to as the plaintext and the encrypted data as ciphertext. The security of encryption lies in the ability of an algorithm to generate ciphertext that is not easily reverted to the original plaintext.

Keys

In the quest for a more secure method of protecting information, the introduction of a *key* adds another level of security. A key is a piece of information that allows only those that hold it to encode and decode a message. Keys come in many different forms such as passwords, numbers generated by an algorithm, digital fingerprints and even electronic devices that work like door keys. It is a series of numbers or symbols that are used to encode a message so that it can only be read by someone in possession of that key or a related key. A key allows both the sender and the recipient of the message to understand how the message has been encrypted and assures them that nobody else knows how it has been encrypted. It is the key that enables the recipient to properly decode the message.

For as long as people have needed to conduct private

conversations across distances, a variety of *encryption* methods have been used to protect secret communications. However, the introduction of electronic communications networks raised an old encryption problem to a new level — how do two people establish secure communications when they live so far apart that they can't meet first to exchange a secret decryption key?

The solution to this problem is an encryption method called Public Key Cryptography (PKC), an ingenious mathematical structure that doesn't need participants to meet beforehand to conduct secure communications. PKC technology also provides a clever method of attaching an encrypted signature to a message to guarantee authenticity. PKC is now the engine for almost all high grade encryption on the *Internet*, including *financial transactions* on web sites and implementation of the highest level *virtual private networks*.

PUBLIC KEY CRYPTOGRAPHY

Public Key Cryptography (PKC) is a near magical property of information arising from the underlying mathematical structure of the universe that also conveniently enables creation of modern-day secure communication channels on the *Internet*.

The main feature of PKC is the use of two keys for each person, a public key and a private key, where either key can decrypt a message encrypted with the other. Each key is almost impossible to find out from the other, and if the keys are long enough the method is effectively unbreakable — according to the known laws of science. The elegant PKC architecture enables clever creation of a secure communications system for distributed participants, which is exactly what is needed for the Internet. The technology is the basis of the field of Public Key Infrastructure (PKI).

COMMON USES OF ENCRYPTION

Authentication

Authentication is the process of logging in, signing on or otherwise presenting information or oneself in a manner that proves his or her identity. The most common example of authentication is the use of a username and password to gain access to a system, network or web site. The username and password combination is often referred to as a person's credentials and it is frequently sent over networks. Encryption is used to protect these credentials. If no encryption is used to protect the information as it is sent over the network, an attacker could capture those credentials and assume the identity of the originator.

Validation: Fingerprints and Digital Signatures

Validation describes the ability to provide assurance that a sender's identity is true and that a message, document or file has not been modified. Encryption can be used to provide validation by making a digital fingerprint of the information contained within a message. A digital fingerprint is a code that uniquely identifies a file or a message by reflecting the content of the file with tremendous specificity.

The encryption program produces the digital fingerprint by performing a byte-by-byte mathematical analysis of the message. Any attempt to modify the message will change the fingerprint. Comparison between a fingerprint known to be good and one sent to the recipient can indicate whether or not the message has been modified. While a fingerprint can indicate that the message has not been tampered with, it does not assure the recipient of the identity of the sender. For that assurance, the sender can utilize a digital signature.

A digital signature is a piece of information that proves the identity of the sender. It is a digital stamp or personal seal that is made using a private key. A sender can electronically or digitally sign a message and its fingerprint before delivery to a recipient. Upon receiving the message, the recipient verifies this signature, using the public key that the sender has previously communicated, indicating that the sender is the expected person. The recipient can verify the fingerprint of the message. Upon validation, the recipient can be reasonably sure that the message came from a trusted person and that the contents of the message have not been modified. Public Key Cryptography (*PKC*) enables user to send a digital signature across the *Internet* to verify a document actually came from user.

A digital signature is like a normal signature, in that it identifies user uniquely and is difficult to fake. A digital signature can be added to an email, document, or other file that user sends to someone else using PKC, enabling the addressee to verify that the communication actually came from the user.

* **Encrypt message :** As normal, first the sender encrypts the message with the recipient's public key.
* **Add signature :** Then the sender adds their signature to the encrypted message, perhaps some text like "This message is from Carol", and then encrypts the whole thing with their own *private* key.
* **Decrypt signature :** The recipient receives the message and decrypts it with the sender's public key, which produces the digital signature and the encrypted message.
* **Decrypt message :** The recipient then decrypts the remaining message with their own private key.

If this unwrapping procedure works, revealing a legible digital signature and a legible message, then the recipient can be sure that the message was sent by the sender, since only they are in possession of the private key used to encrypt the entire message. In the late 1990's and early part of the 2000's, more and more countries started passing legislation recognizing the legal validity of digital signatures in order to provide a foundation for adoption of business on the Internet. A number of companies even established businesses to provide digital signatures to other online business, such as *Verisign*.

Data Protection

Probably the most widely-used application of encryption is in the area of data protection. The information that a business owns is invaluable to its productive operation; consequently, the protection of this information is paramount. For people working in small offices and home offices, the most practical uses of encryption for data protection are file and email encryption.

Encryption of files protects the data that is written to the hard disk on the computer. This information protection is vital in the event of theft of the computer itself or if an attacker successfully breaks into the system. However, file encryption becomes more difficult to use and manage if the office has multiple employees. Because each employee needs the encryption key, protection of the key becomes a more difficult task. The more people who have access to encryption keys, the less effective encryption become. The risk of loss, theft or compromise of information rises as the number of users increases. Files that have been encrypted are also vulnerable to employees who leave the organization or who are disgruntled and may want to cause the organization harm.

Email encryption can be used more easily in office environments as private encryption keys are not generally shared among users and each user has a separate mailbox. When sending a message to multiple recipients, it can be encrypted for each person individually. The encryption key is therefore still private to the sender.

Encryption for E-Commerce: Secure Socket Layers

It is also a valuable security resource for web-based information exchange. The small office/ home office or personal computer user often sees this when doing business via web sites. E-commerce web sites use SSL (Secure Socket Layers) to protect important information such as credit card numbers as they travel across the network. SSL creates a private communication path between the web browser and the web server, encrypting all information that goes between the systems. Most common web browsers have SSL support built in and e-commerce companies can purchase or get freely available web servers that support SSL.

COMPUTER VIRUSES

Computer viruses of one kind or another have infected the Internet since its very first years of existence. Virus protection is now required technology for everyone that uses the Internet. Some viruses are only annoying, displaying a message, using extra memory or disk, or changing file names. However, some are destructive and will change files and erase data, and some will erase the entire hard drive. Some run silently in the background and give outside agents complete control of the computer without the knowledge whenever user are connected to the Internet. The Internet gives viruses a particularly efficient new path for global infection. Some email viruses have spread around the world and brought down tens of thousands of computers in just a few hours. It is absolutely essential that user run an anti-virus protection program to safeguard the computer from these serious threats.

Boot & Program Viruses - Types and Habitats

Boot and program viruses were the first viruses. They are generally made of executable code that hides inside device boot programs and application programs, and are usually targeted for a specific computer operating system. These were the earliest types of computer viruses developed, and remained relatively common in the wild until overtaken in 1998 by script and macro viruses.

Boot viruses : Boot viruses hide in the boot code for a media device, such as a disk or CD, and run automatically when the media is loaded since boot programs are always the first code loaded from any device. Boot viruses proliferated on floppy disks and even CD's into the late 1990's, but is not seen as often these days with the decline in importance of transferable, bootable media.

Program viruses : Program viruses can travel on media

like a CD or across the Internet by email attachment. They hide in an apparently useful program and then run when the program is opened. They are often called Trojan horse viruses, after the hollow wooden horse containing soldiers that Ulysses and the Greeks gave to Minerva during the Trojan war, and from which the soldiers emerged that night to open the gates of the city of Troy to the Greek armies, thereby causing the city's downfall.

Program viruses may be deliberately hidden in a program by the developer, or surreptitiously attached after the fact at some point along its travels from computer to computer. Program viruses are also sometimes the vector of infection for boot viruses and worms.

Virus infection : A greeting card program emailed to user from a friend might display a holiday animation and song, while at the same time installing a remote access virus program that gives a distant hacker control over the computer whenever user is connected to the Internet. Similarly, a shareware program downloaded and emailed to user by another friend might have been infected with a virus on his computer or the server where it was stored. The first thing a boot or program virus often does is insert commands and settings in the operating system so that they can operate freely, undetected, and unaudited, without warning messages or access log records. Some of them even change the Basic Input Output System (BIOS) that interfaces between the computer's hardware and software to help mask their activities.

The most sophisticated program viruses include "stealth viruses", which encrypt their contents to try and avoid detection by virus protection software, and "polymorphic viruses", which alter their content every time they replicate to try and avoid detection, which exhibits behavior just like real viruses. Most anti-virus programs can still catch most of these types of viruses.

Script & Macro Viruses

Script viruses (sometimes called macro viruses) generally travel embedded in email and office automation documents, although they can be found in web pages as well. Old fashioned program viruses are usually implemented in executable system code, whereas script viruses are usually written in a powerful high-level language that is compiled and run on the fly. They often have sophisticated functionality and direct interfaces to high level applications such as word processing, spreadsheet, email, and web programs, and can wreak considerable havoc. Since they first surfaced in office automation programs, they are sometimes also called "macro" viruses.

On computers, turning on the script checking virus protection is essential. However, it should be kept in mind that there may be an associated performance hit for some applications. Many applications on Windows are written in Visual Basic and real-time script virus checking can double the time it takes for their usual functions to run. If user finds that ordinary functions take an inordinate length of time to complete, user can try temporarily turning this feature off in the anti-virus checker — but should not forget to turn it back on afterwards.

Active threats : The following types of script viruses are currently the most active and dangerous, on the Windows platform:

* **Visual Basic** : Visual Basic is a flexible and powerful programming environment for Microsoft Windows, Office, and Internet applications. Script viruses written in Visual Basic can run throughout the architecture, giving them considerable reach and power, and making them the primary virus threat today. The first widespread Visual Basic script virus was Melissa, which brought down several of the large international corporations for several days in March 1999. Melissa traveled in a Microsoft Word document and ran when the document was opened, then opened the associated Microsoft Outlook email program, read the user's email address book, and then sent email copies of it to the first fifty names it found. It spread very quickly. The Melissa virus architecture was quickly

followed by many similar variants programmed by hackers around the world, including the ground breaking Kak, the first Visual Basic script virus that triggered as soon as an email was opened. KAK was then followed by BubbleBoy, which triggered if an email was even viewed in the preview pane. Steady stream of Visual Basic script viruses continue to circulate to this day.

* **ActiveX :** ActiveX is one of Microsoft's distributed application technologies that enable web pages to download programs on the fly with the full power of any executable running on the machine. This makes ActiveX modules especially efficient and powerful, but also a security risk since they can create, change, and deletes files, add system programming code, or take any other action the user account is allowed on the computer.

To help mitigate the risk, Microsoft provides network architecture of encrypted security certificates for ActiveX modules. This network gives user the option of refusing the download of unsigned ActiveX modules from unknown authors, and at least disclosing the signed identity of those modules that user do accept in case they later cause problems.

Hypothetical threats : The following script viruses are largely theoretical, but illustrate that they can turn up wherever there is scripting code:

* *Java* is a standard cross platform development environment, and is often used to download scripts to add functionality like a clock or chat room interface to a web page. Java was written with a strong security model which protects the computer's data and resources, and it has so far proved remarkably resistant to script virus infection. User can turn Java off in the browser if user want to be extra careful, but it will disable some useful functionality on some web pages.
* *JavaScript* is the standard web programming language. JavaScript also has a well-defined security model that

protects data and resources, and the few JavaScript viruses that have been discovered have been mainly theoretical in nature. User can turn JavaScript off in the browser settings if user wants to be extra careful, but it will disable functionality on many web pages.

* *MIME* : The first script virus that triggered as soon as an email was opened was a MIME virus that applied to older versions of Netscape Mail, Microsoft Outlook, and Eudora Mail. In a variation on an old hacker technique, the attached MIME file was given a very long name that triggered a bug which allowed the end of the name to be run as a series of instructions, which could then be written to run the virus. However, a fix for the bug was quickly developed for each vulnerable email program, and MIME viruses have so far remained hypothetical.

* ***Others*** **:** Several other scripting environments have also had viruses, including Corel Draw, Hypertext Preprocessor, Windows Help, Windows installation files, and Windows registry files. Anywhere there is a script interpreter there is an opportunity for a script virus to run.

Internet Worms

Internet worms are truly autonomous virtual viruses, spreading across the net, breaking into computers, and replicating without human assistance and usually without human knowledge. Worms are particularly interesting technological constructs, with an intriguing mathematical structure and complexity. They fascinate because they take the digital imitation of life to another step — they autonomously search for computers, penetrate them, and replicate their intelligence to continue the process. An Internet worm can be contained in any kind of virus, *program* or *script*. Sometimes their inventor will release them into the wild in a single copy, leaving them to replicate by themselves through a variety of stratagems and protocols.

Worms use a variety of methods to propagate across the Internet. Early worms simply scanned the local network drives and folders and inserted themselves into programs wherever they could, trusting human beings to move disks and directories around in the normal course of things so they could continue to spread.

The most successful Internet worm of all time, in terms of sheer saturation, was the *code red* worm, which scanned the Internet for vulnerable Windows computers running the IIS web server to install it and continue the infection.

A wide range of other inventive strains of Internet worms have employed security weaknesses in IRC, MAP1, sendmail, finger, and other programs and protocols. A few worms began to be discovered for Linux in the late 1990's as it became more popular across the Internet and some vulnerabilities were found, but the strong security architecture of Linux has kept the number of problems relatively low.

The first worm : The first worm disabled most of the Internet then existing. Robert Morris, a Computer Science graduate student at Cornell University and (embarrassingly) son of the Chief Scientist at the National Computer Security Center, wrote a 99 line program in the C language designed to *self-replicate* and propagate itself from machine to machine across the Internet. The worm performed the trick by combining a bug in the debugging mode of the sendmail program used to control email on almost all Internet computers, a bug in the *finger* program, and the *Unix* rexec and rsh commands.

On November 2, 1988, Morris released his worm, but did so from an MIT computer to disguise his origin. In his view, only one thing went wrong — the worm started replicating at a much faster rate than he had predicted, and began *crashing* and disabling computers across the Internet. Morris sent out an anonymous message telling people how to disable the worm, but because it had brought down the Internet, the message about how to disable it couldn't get through. The worm eventually infected more than 6,000 computers across the Internet. Within a day teams of programmers at the University of California at Berkeley and Purdue University reverse engineered the worm and developed methods of stopping it. The Internet then came back to normal in a couple of days.

Virus Protection

The most important computing advice is "back up the files", which helps to safeguard the data if user ever gets a virus. The second most important principle is "run an anti-virus protection program". If the anti-virus program does not include a good *firewall*, user must obtain one of those as well. Modern computer viruses are more virulent than ever. It is critically essential for the protection of all of the valuable programs and information on the computer that user run a good anti-virus protection program. Most of these applications can regularly update their database over the *Internet* as the threats evolve and automatically keep the anti-virus protection up-to-date and the computer safe.

* **Free** : A few companies offer a free anti-virus protection option:

  AVG Anti-Virus

  OpenAntiVirus

* **Indexes** : The following indexes list companies that provide anti-virus protection programs and services:

  Google - Anti-Virus

  Freebyte's Guide to free Anti-Virus Software

  Yahoo - Virus Protection

* **Commercial** : The following companies are leading anti-virus protection providers:

  Avira.com

  BitDefender.com

  CentralCommand.com

  Dr. Solomon's Anti-Virus Toolkit

  F-Prot Antivirus

F-Secure Anti-Virus

Kapersky Labs

McAfee Associates' VirusScan

NOD32 Ireland

Panda Software Anti-Virus

Sophos Anti-Virus

Symantec's Norton Anti-Virus

TrendMicro.com

Maintenance : Once user has installed anti-virus protection, he should take the following additional protective measures:

Never use a floppy disk, CD, DVD, tape, or other external media that has been on someone else's computer without first scanning it with the anti-virus protection program, which should be set to scan all media by default. If user lends media to someone else to copy a file, write-protect it first so that it won't get inadvertently infected.

Protect the perimeter : Make sure the anti-virus protection settings are turned on by default to scan files incoming over email and downloaded off the Internet.

Infection : Computers that run good anti-virus protection usually don't get infected. However, if user is sure that the system has somehow got a virus anyway, user can take the following steps:

* Immediately shutdown the computer, and do not reboot it from the infected disk, in order to prevent the virus from wreaking more damage.
* Boot the computer from some clean external media such as a bootable floppy, CD, DVD, or external disk that has previously been scanned by the anti-virus protection.
* Run the anti-virus protection software from the clean boot disk, on the infected disk, and if required fix or delete infected files and replace them on the infected disk.
* If user needs help or the anti-virus protection can't clean

the disk, then user are best advised to take the computer to a good professional repair shop where they have tools to try and clean and recover the disk as best as possible.

* Keeping in mind that anti-virus protection sometimes generates false alarms — a common cause is when a program file has changed size but for a valid reason. Another common indicator that user may have a false alarm is if the anti-virus protection claims that a file may contains a virus but doesn't know the virus's name.

Virus Hoaxes

The first rule of all email alerts, especially virus alerts, is that the most urgent and seemingly authentic warnings are the ones most likely to be a hoax. They can be forwarded to so many people at once so easily, it only takes a small percentage of recipients to spread the well-meaning but destructive infection farther.

Fake alerts go well beyond virus hoaxes, and include any email warning about consumer alerts, health or safety warnings, missing children, impending draconian legislation, or other apparently alarming issue. Any virus or other email alert should be considered a hoax unless specifically verified true. Hoax email warnings rarely contain valid contact information, and when they do, the inclusion is unauthorized.

A common and telltale sign of a virus hoax or fake alert is a heartfelt appeal at the end of the email to forward this critically important message to everyone user know (without which the writer's efforts will have been in vain).

Not only do virus hoaxes and other fake alerts spread across the world very quickly, they also change and mutate over the weeks and years, with more tantalizing or outrageous details being added along the way, sometimes with the bad guy's names updated to the current villains. Soon hoaxes start showing up on people's web pages, which makes them seem more authentic, and then, when it is occasionally discovered that supposedly true information is outrageously false, our threshold of acceptance for all information edges just a bit higher as we turn up the gain on the filters we use to judge the accuracy of all information from the outside world.

Resources : A couple of well known Internet hoaxes are described in the myths and legends section. The following web sites track virus hoaxes and other fake alerts and maintain searchable databases that should always be consulted to determine an alert's authenticity before passing it on:

1. *Google — Virus Hoaxes*
2. *Yahoo — Urban Legends - Viruses*
3. *CIAC Internet Hoaxes*
4. *Computer Virus Myths Home Page*
5. *F-Secure Hoax Warnings*
6. *Fraud.org*
7. *McAfee Virus Hoaxes*
8. *Sophos Virus Hoaxes*
9. *Symantec Virus Hoaxes*

Architecture of Internet

IP Addresses

Each computer on the *Internet* has a unique numerical address, called an Internet Protocol (IP) address, used to route packets to it across the Internet. Just as the postal address enables the postal system to send mail to the house from anywhere around the world, the computer's IP address gives the Internet *routing* protocols the unique information they need to route *packets* of information to the desktop from anywhere across the Internet. If a machine needs to contact another by a *domain name*, it first looks up the corresponding IP address with the *domain name service*. The IP address is the geographical descriptor of the virtual world, and the addresses of both source and destination systems are stored in the header of every packet that flows across the Internet. User can find the IP address on a Windows computer by opening an MSDOS or Command window and typing one of "winipcfg" or "ipconfig". User can find the IP address on a Mac computer by checking the Network control panel. No matter what electronic device user are using and where user are, if user are connected to the web user can visit the following sites to dynamically find the IP address in real time:

* *IP Info*
* *IP2Location.com* (time intensive)
* *Network-Tools.com* (time intensive)
* *ShowIPAddress.com*
* *Stealthtests*

1. Format : An IP address is made up of four bytes of information (totaling 32 bits) expressed as four numbers between 0 and 255 shown separated by periods. For example, the computer's IP address might be 238.17.159.4, which is shown below in human-readable decimal form and in the binary form used on the Internet.

Example IP Address

Decimal: 238 . 17 . 159 . 4

Binary: 11101110 00010001 10011111 00000100

Each of the four numbers uses eight bits of storage, and so can represent any of the 256 numbers in the range between zero (binary 00000000) and 255 (binary 11111111). Therefore, there are more than 4 billion possible different IP addresses in all:

4,294,967,296 = 256 * 256 * 256 * 256

2. Allocations : The Internet Assigned Numbers Authority manages the allocation of IP addresses to different organizations in various sized blocks. The IANA IP Address service page provides a focal point for this world wide IP address management. An official list of the allocations of IP address blocks can be found at the Internet Protocol Address Space site, and related information can be found at the IP Index Encyclopedia. Most of the address blocks have been allocated to research, education, government, corporations, and Internet

Service Providers, who in turn assign them to the individual computers under their control. A few addresses are reserved for future or special use.

If user connects to the Internet over a phone line, then the IP address is probably assigned dynamically by the Internet service provider from an available pool of addresses each time user log on. If the computer is permanently connected to an Internet network, such as at the office or on a high speed home connection, then the IP address could be permanently assigned, or could be reassigned each time user reboot the computer.

3. Lookup databases : User can find out more information on any particular IP number by searching one of the following databases. Since each address is generally kept in only one database, user might have to try more than one to find the database that has the information on any particular address:

* American Registry for Inernet Nembers
* European IP Address Allocations
* Asia Pacific IP Address Allocations; including country list

DOMAIN NAME SYSTEM (DNS)

The Domain Name System (DNS) as a whole consists of a network of servers that map Internet domain names like www.google.com to a local IP Addresses. The DNS enables domain names to stay constant while the underlying network topology and IP addresses change. This provides stability at the application level while enabling network applications to find and communicate with each other using the Internet Protocol no matter how the underlying physical network changes.

INTERNET DOMAIN NAMES

Internet domain names come in four main types:

* top-level domains,
* second-level domains,
* third-level domains, and

* country domains.

Internet domain names are the alphanumeric identifiers we use to refer to hosts on the Internet, like "google.com". The structure of Internet domain names was first described in RFC 819 (Request for Comments RFC documents are a series of memoranda encompassing new research, innovations, and methodologies applicable to Internet technologies), and their syntax was first described in RFC 822. User can look up existing Internet domain names with the whois service, and get the own domain name from an accredited registrar.

1. Top-level domains : Internet domain names are organized by their levels, with the higher levels on the right. For example, for the domain "mail.twenty.net" the top-level domain is "net", the second-level domain is "twenty.net", and the third-level domain is "www.twenty.net". Originally, the top-level Internet domains were associated with organizations, mainly "*.arpa*", "*.csnet*", "*.bitnet*", and "*.uucp*". On January 28, 1986, coincidentally the same day the *Challenger space shuttle* exploded, those same four organizations met at *SRI* and agreed to restructure the domain name space into the following eight subject-specific top-level domains, with technical administration performed by Information Science Institute (ISI) :

TLD	*Usage*
.bitnet	Used for computers on the *BITNET* network.
.com	Originally stood for "commercial" to indicate a site used for commercial purposes, but it has since become the most well-known top-level Internet domain, and is now used for any kind of site.
.int	Used by "International" sites, usually NATO sites.
.edu	Used for educational institutions like universities.
.gov	Used for US Government sites.
.mil	Used for US Military sites.
.net	Originally intended for sites related to the Internet itself, but now used for a wide variety of sites.

.org	Originally intended for non-commercial "organizations", but now used for a wide variety of sites. Managed by the Internet Society.

In December 2002, after a long debate, ICANN (Internet Corporation for Assigned Names and Numbers) approved the following additional top-level Internet domains: .aero, .biz, .coop, .info, .museum, .name, and .pro. In October, 2004, ICANN approved the top-level domains .travel and .post. Periodically new top level domains continue to be approved.

2. Second-level domains : Top-level Internet domains like ".com" are shared by all the organizations in the domain. Second-level domain names like "yahoo.com" and "livinginternet.com" are registered by individuals and organizations. Second-level domains are the addresses commonly used to host Internet applications like web hosting and email addressing. Excluding the top-level domain portion, second-level domain names can have up to 61 characters. For many years, characters were restricted to the 26 letters, 10 numbers, or the hyphen character, except the hyphen can't be the first or last character. Under these conditions, there are 36 possibilities for the first and last character of the domain name, and 37 possibilities for the other 59 characters. Therefore, the total number of possible different second level domain names was:

37^59 x 36 x 36 = 433,018,769,429,481, 000,000,000,000, 000,000,000,000,000,000, 000,000,000, 000,000,000,000, 000,000,000, 000,000, 000,000, 000,000, 000

3. Third-level domains : Third-level Internet domain names are created by those that own second-level domains. Third-level domains can be used to set up individual domains for specific purposes, such as a domain for web access and one for mail, or a separate site for a special purpose.

4. Fourth-level and even higher Internet domains : Fourth-level and even higher Internet domains like "www2.un.news.media.twenty.net" can be also be established.

Three of four levels are usually sufficient for most purposes.

Country domains : Each country in the world has its own top-level Internet domain with a unique alphabetic designation. A few countries and example domains are shown below.

Top-Level Domain	*Country*	*Example site*
.ae	United Arab Emirates	*http://www.uae.org.ae/*
.at	Austria	*http://www.austria.at/*
.au	Australia	*http://www.morning.com.au/*
.br	Brazil	*http://home.iis.com.br/*
.ca	Canada	*http://www.canoe.ca/*
.ch	Switzerland	*http://www.badminton.ch*
.us	United States	*http://www.co.maui.hi.us/*

Organizations in each country are responsible for managing the top-level Internet domain, and then for allocating second-level domains within that domain to people and organizations with interests in that country. The Country Code Domain Name Supporting Organization (CCNSO) is a support organization of ICANN responsible for developing consensus positions and recommending global policies relating to country-code top-level domains.

DOMAIN NAME SERVERS

The Domain Name System (DNS) servers distribute the job of mapping domain names to IP address among servers allocated to each domain. Each second-level domain must have at least one domain name server responsible for maintenance of information about that domain and all subsidiary domains, and response to queries about those domains from other computers on the Internet. For example, management of domain name information and queries for the yahoo.com domain is handled by a specific DNS server that takes care of the load required. This distributed architecture is designed to enable the Internet to grow, where as the number of domains grew, the number of DNS servers can grow to keep pace with the load. Today, everyone who registers a second-level domain name must at the same time designate two DNS servers to manage queries and return the current IP address for addresses in that domain. The primary domain name server is always consulted first, and the secondary domain name server is queried if the primary doesn't answer, providing a backup and important support to overall Internet reliability.

The application that underlies almost all DNS server software on the Internet is an open source program called BIND (An open source program for Domain Name Services. Software like Bind is necessary for computers to translate text Web address (e.g. www.yahoo.com) into the numeric IP (Internet Protocol) addresses they need to retrieve Web pages), currently maintained by the Internet Systems Consortium. When the computer was added to the Internet, one of the initial setup tasks was to specify a default domain name server, usually maintained by the local Internet Service Provider, and almost certainly a variant of the BIND server software. The IP address assigned to a computer may change frequently because of physical moves or network reconfigurations. The major advantage of the network of DNS servers is that domain names stay the same even when IP addresses change, and so the domain name servers can transparently take care of the mapping. If user runs DNS server software, user should always make sure it is up-to-date with the latest version and patches. DNS server vulnerabilities typically affect the systems running the servers, which is generally Internet Service Providers, and so are not a direct threat to the home user unless user are running one at home.

INTERNET ROUTING

IP Addresses and packet switching provide the technical infrastructure which routing protocols use to transmit packets across the Internet. The Internet Protocol transfers packets between networks and provides the software bridge that knits the whole thing together.

INTERNET ROUTERS

Internet routers are specialized computers that interconnect the network by switching communications from one line to another at cross points. When a computer communicates with another on the Internet, it addresses each packet with the other computer's IP address and then sends it to the closest Internet router. The router then uses a routing algorithm to send the packet across the Internet to the destination computer.

An Internet router is typically connected to several different networks. An organization building often has several routers connecting to various local area networks, found in "wiring closets" on each floor jammed full of network equipment and cables. Each organization building typically has one router for connection to the wide area Internet itself.

* **Interior Gateway Protocols (IGP) :** Interior Gateway Protocols (IGP) are used to route Internet communications within a local area network, such as within an office building.
* **Exterior Gateway Protocols (EGP) :** While IGP is used within local networks, Exterior Gateway Protocols (EGP) are used for routing between networks, generally on the Internet backbone itself, linking the different networks together.

TCP/IP model

The *TCP/IP* **model** or **Internet reference model**, sometimes called the **DoD model** (DoD, Department of Defence), ARPANET reference model, is a layered abstract description for communications and computer network protocol design. It was created in the by DARPA in 1970s for use in developing the Internet protocols. The structure of the Internet is still closely reflected by the TCP/IP model. The original TCP/IP reference model was consisted of 4 layers, but is now viewed by some as a 5-layer model, even though no IETF (**Internet Engineering Task Force**) standards-track document has accepted a five-layer model, and IETF documents indeed deprecate strict layering of all sorts. Due to the lack of acceptance of the five-layer model by the body with technical responsibility for the protocol suite, it is not unreasonable to regard five-layer presentations as teaching aids, possibly to make the IP suite architecture more familiar to those students who were first exposed to OSI layering. This model was developed before the OSI Reference Model and the IETF.

OSI AND TCP/IP LAYERING DIFFERENCES

The three top layers in the OSI model - the *application layer*, the *presentation layer* and the *session layer* - usually are lumped into one layer in the TCP/IP model. While some pure OSI protocol applications, such as *X.400*, also lumped them together, there is no *requirement* that a TCP/IP protocol stack needs to be monolithic above the transport layer. For example, the *Network File System* (NFS) application protocol runs over the *eXternal Data Representation* (XDR) presentation protocol, which, in turn, runs over a protocol with session layer functionality, *Remote Procedure Call* (RPC). RPC provides reliable record transmission, so it can run safely over the best-effort *User Datagram Protocol* (UDP) transport. The session layer roughly corresponds to the Telnet *virtual terminal* functionality, which is part of text based protocols such as *HTTP* and *SMTP* TCP/IP model application layer protocols. It also corresponds to TCP and UDP port numbering, which is considered as part of the transport layer in the TCP/IP model. The presentation layer has similarities to the *MIME* standard, which also is used in *HTTP* and *SMTP*. The associated OSI development effort, however, has been abandoned given the real-world adoption of TCP/IP protocols.

7	**Application**	*ECHO, ENRP, FTP, Gopher, HTTP, NFS, RTSP, SIP, SMTP, SNMP, SSH, Telnet, Whois, XMPP*
6	**Presentation**	*XDR, ASN.1, SMB, AFP, NCP*
5	**Session**	*ASAP, TLS, SSL,* ISO 8327 / CCITT X.225, *RPC, NetBIOS, ASP*
4	**Transport**	*TCP, UDP, RTP, SCTP, SPX, ATP, IL*

3	**Network**	*IP, ICMP, IGMP, IPX, OSPF, RIP, IGRP, EIGRP, ARP, RARP, X.25*
2	**Data Link**	*Ethernet, Token ring, HDLC, Frame relay, ISDN, ATM, 802.11 WiFi, FDDI, PPP*
1	**Physical**	*10BASE-T, 100BASE-T, 1000BASE-T, SONET/SDH, G.709, T-carrier/E-carrier*, various *802.11* physical layers

VARIOUS LAYERS AND PROTOCOLS

The layers

The following is a description of each layer in the IP suite stack:

1. Application layer

The *application layer* is used by most programs for network communication. Data is passed from the program in an application-specific format, then *encapsulated* into a transport layer protocol

Since the IP stack has no layers between the application and transport layers, the application layer must include any protocols that act like the OSI's presentation and session layer protocols. This is usually done through *libraries*. Data sent over the network is passed into the application layer where it is encapsulated into the application layer protocol. From there, the data is passed down into the *lower layer protocol* of the transport layer.

The two most common lower layer protocols are TCP and UDP. Common *servers* have specific *ports* assigned to them while *clients* use *ephemeral ports.*

Routers and *switches* do not utilize this layer but *bandwidth throttling* applications do, as with the *Resource Reservation Protocol* (RSVP).

2. Transport layer

The *transport layer*'s responsibilities include end-to-end message transfer capabilities independent of the underlying network, along with error control, fragmentation and flow control. End to end message transmission or connecting applications at the transport layer can be categorized as either:

connection-oriented e.g. *TCP*

connectionless e.g *UDP*

The transport layer can be thought of literally as a transport mechanism e.g. a vehicle whose responsibility is to make sure that its contents (passengers/goods) reach its destination safely and soundly, unless a higher or lower layer is responsible for safe delivery. Some applications, such as *Voice Over IP* (VOIP) can tolerate dropped packets, but not delay or reordering that would be caused by a reliable transport.

The transport layer provides this service of connecting applications together through the use of *ports*. Since IP provides only a *best effort delivery*, the transport layer is the first layer of the TCP/IP stack to offer reliability. *TCP* is a connection-oriented protocol that addresses numerous reliability issues to provide a *reliable byte stream*:

* data arrives in-order
* data has minimal error (i.e correctness)
* duplicate data is discarded
* lost/discarded packets are resent
* includes traffic congestion control

Both TCP and UDP are used to carry a number of higher-level applications. The applications at any given network address are distinguished by their TCP or UDP *port*.

3. Network layer

As originally defined, the *Network layer* solves the problem of getting packets across a single network. Examples of such protocols are *X.25*, and the *ARPANET*'s *Host/IMP Protocol*. With the advent of the concept of *internetworking*, additional functionality was added to this layer, namely getting data from the source *network* to the destination network. This generally involves routing the packet across a network of networks, known as an internetwork or (lower-case) internet.

In the Internet protocol suite, *IP* performs the basic task of getting packets of data from source to destination. IP can carry data for a number of different *upper layer protocols*; these protocols are each identified by a unique protocol number: *ICMP* and *IGMP* are protocols 1 and 2, respectively. What makes them part of the network layer is that their payload is totally concerned with management of the network layer.

4. Data link layer

The link layer, which is the method used to move packets from the network layer on two different hosts, is not really part of the Internet protocol suite, because IP can run over a variety of different link layers. The processes of transmitting packets on a given link layer and receiving packets from a given link layer can be controlled both in the *software device driver* for the *network card*, as well as on *firmware* or specialist *chipsets*. These will perform *data link* functions such as adding a *packet header* to prepare it for transmission, then actually transmit the frame over a *physical medium*.

The link layer can also be the layer where packets are intercepted to be sent over a *virtual private network*. When this is done, the link layer data is considered the application data and proceeds back down the IP stack for actual transmission. On the receiving end, the data goes up the IP stack twice (once for routing and the second time for the VPN).

The link layer can also be considered to include the physical layer, which is made up of the actual physical network components (hubs, repeaters, fiber optic cable, coaxial cable, network cards, Host Bus Adapter cards and the associated network connectors: RJ-45, BNC, etc), and the low level specifications for the signals (voltage levels, frequencies, etc).

5. Physical layer

The Physical layer is responsible for encoding and transmission of data over network communications media. It operates with data in the form of bits that are sent from the Physical layer of the sending (source) device and received at the Physical layer of the destination device.

Ethernet, Token Ring, SCSI, hubs, repeaters, cables and connectors are standard network devices that function at the Physical layer. The Physical layer is also considered the domain of many hardware-related network design issues, such as LAN and WAN topology and wireless technology.

Internet protocol suite

The **Internet protocol suite** is the set of *communications protocols* that implement the *protocol stack* on which the *Internet* and most commercial networks run. It has also been referred to as the TCP/IP *protocol suite*, which is named after two of the most important protocols in it: the *Transmission Control Protocol* (TCP) and the *Internet Protocol* (IP), which were also the first two networking protocols defined.

The Internet protocol suite — like many protocol suites — can be viewed as a set of layers. Each layer solves a set of problems involving the transmission of data, and provides a well-defined service to the *upper layer protocols* based on using services from some lower layers. Upper layers are logically closer to the user and deal with more abstract data, relying on *lower layer protocols* to translate data into forms that can eventually be physically transmitted

The *OSI model* describes a fixed, seven-layer stack for networking protocols. Comparisons between the OSI model and TCP/IP can give further insight into the significance of the components of the IP suite. Both the OSI and the TCP/IP models are 'standards' and application developers will often implement solutions without strict adherence to proposed 'division' of labour within the standard whilst providing for functionality within the application suite. This separation of 'practice' from theory often leads to confusion, although it may also lead to more efficient implementation.

REFERENCES

1. *Some Internet Architectural Guidelines and Philosophy,RFC 3439*, R. Bush & D. Meyer, *December 2002*
2. Atkin, D. J., Jeffres, L. W., & Neuendorf, K. A. (1998). Understanding Internet adoption as telecommunications behavior. *Journal of Broadcasting & Electronic Media*, **42**(4), 475-490.
3. Australian Bureau of Statistics. (1997). *Australian standard classifications of occupations.* (2nd Edition). Canberra, Australia: Australian Bureau of Statistics..
4. *Broadband killed the television star.* (2004, May 24). *The Register* Retrieved 25 March, 2005 from http://www.theregister.co.uk/2004/05/27/broadband_threatens_tv/.
5. Bromley, R. & Bowles, D. (1995). Impact of Internet on use of traditional news media. *Newspaper Research Journal*, **16**(2), 14-27.
6. Brown, C. (1999, June). *State of the American newspaper. FEAR.COM. American Journalism Review*, http://www.ajr.org/Article.asp?id=3230.
7. Case, Donald. (2002). *Looking for information. A survey of research on information seeking, needs and behavior.* San Diego, CA: Academic Press.
8. Chen, W., Boase, J. & Wellman, B. (2002). The global villagers: comparing Internet users and uses around the world. In Barry Wellman & Carolyne Haythorntwaite (Eds), *The Internet in everyday life* (pp. 74-113). Oxford: Blackwell.
9. Dupagne, M. (1994). Testing the relative constancy of mass media expenditures in the United Kingdom. *The Journal of Media Economics*, **7**(3), 1-14.
10. Dupagne, M. (1997). Beyond the Principle of Relative Constancy: determinants of consumer mass media expenditures in Belgium. *The Journal of Media Economics*, **10**(2), 3-19.
11. Dutta-Bergman, M. (2004). Complementarity in consumption of news types across traditional and new media. *Journal of Broadcasting and Electronic Media,* **48**(1), 41-61.
12. Hsia, H. (1989). Introduction. In Jerry Salvaggio and Jennings Bryant (Eds.), *Media use in the information age* (pp xv-xxviii). Mahwah, NJ: Lawrence Erlbaum Associates.
13. James, M., Wotring, C. & Forrest, E. (1995). An exploratory study of the perceived benefits of electronic bulletin board use and their impact on other communication activities. *Journal of Broadcasting and Electronic Media,* **39**(1), 30-50.
14. Kayany, J. & Yelsma, P. (2000). *Displacement effects of online media in the socio-technical contexts of households. Journal of Broadcasting and Electronic Media,* **44**(2), 215-229.

15. Lacy, S. & Noh, G. (1997). Theory, economics, measurement, and the Principle of Relative Constancy. *The Journal of Media Economics,* **10**(3), 3-16.

16. LaRose, R., & Eastin, M. (2004). A social cognitive theory of Internet uses and gratifications: toward a new model of media attendance. *Journal of Broadcasting & Electronic Media,* **48**(3), 358-377.

17. Lee, P. & Leung, L. (2004). *Assessing the displacement effects of the Internet.* Paper presented at the International Conference on Internet Communication in Intelligent Societies. School of Journalism & Communication, Chinese University of Hong Kong, July 8-10.

18. Lent, A. (2000, January 11). *Livin' la Vida Internet. Sure, you've got information close at hand, but what's it do for your quality of life? PC World.* http://www.pcworld.com/news/article/0,aid,14746,00.asp.

19. *www.webopedia.com/TERM/I/Internet.html*

20. *www.webopedia.com/DidYouKnow/Internet/2002/Web_vs_Internet.asp*

21. netvalley.com/intval.html

22. *http://en.wikipedia.org/wiki/Internet*

23. searchwindevelopment.techtarget.com/sDefinition/0,,sid8_gci212370,00.html

24. *www.isoc.org/internet/history*

25. *www.cisco.com/univercd/cc/td/doc/cisintwk/ito_doc/ip.htm*

26. *www.linuxjunkies.org/network/tcpip/intro0.html*

27. *www.rfc-editor.org/rfcxx00.html*

28. *www.dmoz.org/Computers/Internet/Protocols*

29. *www.infosyssec.net/infosyssec/netprot1.htm*

30. www.erg.abdn.ac.uk/users/gorry/course/inet-pages/ip.html

31. www.internet.com

32. Levy, Mark & Windahl, Sven (1989). The concept of audience activity. In K. Rosengren, L. Wenner & P. Palmgreen (Eds.), *Media gratifications research: current perspectives*. (pp. 109-122). Beverly Hills, CA: Sage Publications.

33. Lin, C. (2001). Audience attributes, media supplementation and likely online service adoption. *Mass Communication and Society,* **4**(1), 19-38.

34. *Architectural Principles of the Internet, RFC 1958,* B. Carpenter, *October 1989*

35. *Rethinking the design of the Internet: The end to end arguments vs. the brave new world,*D. Clark & M. Blumenthal, *August 2000*

36. The Internet Protocol Journal Book Review Internetworking with TCP/IP Internetworking with TCP/IP (Vol. 1): Principles, Protocols, and Architectures, Douglas E. Comer, *ISBN 0-13-018380-6,* Prentice Hall, 2000.

37. OSI: Reference Model Addendum 1: Connectionless-mode Transmission, ISO7498/AD1],ISO7498/AD1, May 1986
38. *Internal Organization of the Network Layer*, ISO 8648
39. *Open Systems Interconnection — Basic Reference Model — Part 4: Management framework*,ISO 7498/4
40. *Framework and taxonomy of International Standardized Profiles*, ISO 10000, October 1998
41. *Requirements for Internet Hosts — Communication Layers,RFC 1122*,R. Braden, October 1989
42. Ronda Hauben. *From the ARPANET to the Internet*. TCP Digest (UUCP).
43. Forouzan, Behrouz A. (2003). *TCP/IP Protocol Suite*, 2nd, McGraw-Hill. *ISBN 0-07-246060-1.*
44. *Andrew S. Tanenbaum*. Computer Networks. *ISBN 0-13-066102-3*
45. *Douglas E. Comer*. Internetworking with TCP/IP - Principles, Protocols and Architecture. *ISBN 86-7991-142-9*
46. *Joseph G. Davies* and *Thomas F. Lee*. Microsoft Windows Server 2003 TCP/IP Protocols and Services. *ISBN 0-7356-1291-9*
47. *Craig Hunt* TCP/IP Network Administration. O'Reilly (1998) *ISBN 1-56592-322-7*
48. *W. Richard Stevens*. The Protocols (TCP/IP Illustrated, Volume 1). Addison-Wesley Professional; 1st edition (December 31, 1993). *ISBN 0-201-63346-9*
49. *Ian McLean*. Windows(R) 2000 TCP/IP Black Book *ISBN 1-57610-687-X*
50. *RFC 675* - Specificiation of Internet Transmission Control Program, December 1974 Version

10

Internet : Application to Libraries

Libraries are more than just books on the shelves. Everything is simplified with the help of high-speed Internet. Libraries continue to develop new methods and uses for high-speed internet to keep patrons up-to-date. The speed of the Internet has changed the way people receive information in the library. It combines the immediacy of broadcast with the in-depth coverage of newspapers and making it a perfect source for news and weather information. The services available due to Internet are powerful. Now days library becomes capable of giving access to information from anywhere in the world. The networked services such as CD-ROM networks and word processing facilities remain available apart from Internet services. One of the main reasons most libraries provide access to the Internet is to give their users access to another information resource.

Downloading of information from the web is done either to the computer's hard drive or to floppy disks. Allowing users to download to floppy disks also carries some risks, particularly if users are permitted to use their own disks. For security reasons, many libraries only allow downloading to disks provided by library itself. Copyright applies to inormation on the Internet, and users should be made aware of this. Based on requirement offered libraries need to ensure that their browsers are kept updated with the latest plug-ins such as Acrobat, Flash, Macromedia, VRML etc. Sometimes frequent updating is essential. Apart from this filtering software are necessary. It is now a trend to arrang links to sites, selecting and signposting good sites, and providing subject lists, resource directories, and links.

ASSISTANCE AND TRAINING

Even today, some users may inevitably require help, and consideration should be given as to the level of assistance which will be provided. Librarians have always helped users to find information in books, and this is just another format. However, sometimes, more help is needed because the medium may be new to many users requiring assistance should be available at all times. Libraries could provide introductory help just to get people started and sometimes information library oriented user-education. It is important that users know what they can expect. Libraries will need to consider whether or not to limit access to serious research only. Library should understand te user needs as well. Things to be taken into consideration here include the purpose of the library, the number of terminals available, the mechanics of preventing access to 'unsuitable' sites, and indeed the definition of an 'unsuitable' site.

Major objectives

To provide staff with the opportunity to:

1. Acquire an understanding of the fundamental principles and best practices relevant to providing information services for remote users.
2. Examine and engage with emerging web technologies relevant to a variety of information environments.

User objectives

Upon successful completion of the training, a staff person should be able to:

* Demonstrate understanding of the theory and practice of making information services accessible to remote users.
* Identify major issues and current trends in networked information services.
* Understand, evaluate, and use web technologies and tools relevant to a variety of information environments.
* Design and deploy one or more web technologies and tools relevant to the student's current and projected information environments.
* Develop a personal philosophy based on the profession's standards of best practices regarding the evolving provision of information services to remote users.

APPLICATION TO LIBRARIES

The Internet has rapidly spread around the world in the past decade. Over the past few decades a variety of new and exciting information services have appeared on the Internet, each with its own distinct characteristics. Computer and Communication technologies have revolutionized the whole world of information and have penetrated in areas beyond libraries transforming nearly every facet within society. Powerful PCs, sophisticated network technologies, affordable connectivity have given a boost for Internet. Use of the Internet and other network resources is changing the traditional library functions. While in the 1980's and early 1990's the Internet's modalities were changing traditional library functions and were used mainly for communications, database searching and bibliographic access etc. By *online*, it is meant that users are likely be geographically isolated, studying perhaps from home, but with full access to all of the necessary resources via their Internet connection. Materials (information sources) and processes (such as downloading, uploading, accessing etc.) are electronic, and interaction will be either synchronous (eg via chat-rooms or videoconferencing facilities) or asynchronous (eg via email or discussion lists). The Internet applications are listed as following:

OPAC

OPAC stands for online public access catalogue. The OPAC allows search by author, title, subject, call number, keywords, and Boolean search. The Call Number indicates the location of the book on the self. The computer catalogue, OPAC is extremely user-friendly and enables searches to be conducted with ease and speed. The online catalogue can tell user what is available in the Library, where to find it, and whether it is on loan or not. If all copies are on loan user can reserve it online. The entire stock of the Library is on the catalogue. This may include books, journals, music scores, Media Resources equipment, videos, sound recording and dissertations.

E-mail

The Internet is not purely a source of information however; it is also a means of communication, and many users will want to use it to send and receive messages. It is also possible to send messages through an Internet browser, but although mail can be sent, there is no way to receive a reply as browsers have a general rather than a person-specific e-mail address.

However, e-mail is also a means of obtaining information. And if the library supports lifelong learning it may want to allow learners to contact tutors and fellow students. People using *free* web-based e-mail such at Hotmail or Yahoo mail are simply visiting another web site, so it is difficult to prevent, and there are no technical or financial reasons to do so. They register under their own name with the web site offering the service.

BULLETIN BOARDS AND MAILING LISTS

These allow users to place messages on the Internet which are delivered to all subscribers via e-mail. Recipients can then reply in their own time. They are a very useful source of information and providing access. Of course, as these require access to e-mail, they will be unavailable if e-mail is not a service offered. The rapid growth of information, coupled with the ability to exchange it more rapidly among more people than ever before, is creating a new environment for education.

CHAT ROOMS

Chat rooms are very similar to bulletin boards and mailing lists except that they operate in real time. Messages are received and responded to by people who are online in the chat room at the same time. Like bulletin boards and mailing lists they can be a very useful source of information where people with mutual legitimate interests can exchange ideas.

TELNET

Every network manager has used a telnet client at one time or another. The telnet specifications were set in 1983 as part of RFC 854. Telnet is a client-server application, which means that one machine has to provide the Telnet service (or in UNIX terms, the Telnet daemon). Telnet allows one to become a user on a remote computer. Telnet uses TCP, an OSI layer 4 (or transport layer) connection oriented protocol who's primary function is to make sure that each and every packet arrives to it's destination by way of acknowledgement frames. Now that most resources can be viewed using a web browser, Telnet is not as essential as it once was for searching purposes. However, there are still some sites, mostly library catalogues as it happens, which can only be accessed via Telnet, and Telnet is also necessary for users who wish to use a library workstation to access e-mail on a server elsewhere.

E-commerce

This generally means buying and selling goods or services over the Internet, and might be regarded by some libraries as inappropriate use of a library's computer. However, some libraries might want to provide their users with access to online bookshops; one can see Essex Libraries' online shopping arcade which links users to amazon.com and WH Smith where they can purchase books, music etc.. Business use corresponding to the type of service traditionally offered by library business information services is of course acceptable, and indeed desirable, contributing as it does to the local economy.

INTERACTIVE LEARNING SERVICES

Libraries have always been in the education business, and are increasingly offering flexible or open learning services as part of the lifelong learning. It is eminently desirable therefore to consider offering access to online information sources, not only on such subjects as how to use the library and Internet, but on a range of other subjects too. Useful links to suitable courses can be installed on the library's computers, and it is particularly useful to have Internet tutorials signposted from the home page for the benefit of novice users.

Rapid technological advancement may produce problems and challenges for educational institutions when their products and services are rendered obsolete virtually within a short time-horizon. The librarian who has properly learned his/her craft will have transferable skills and knowledge perfectly adaptable to the emerging technology. The benefit of having transferable knowledge in such a volatile marketplace is readily apparent. It is insurance for survival of the Web- based sources. The Web sites have high dynamic rates of birth and death. The Internet is a graveyard of Web sites who tried but failed to keep up with the contents that the visitors really need from them. Many got on the Web very quickly once it was clear that many new sites were choked with flow and did not have any useful and interesting information. There is certainly a power in the Internet communication, development, and delivery of intellectual materials via this medium we are mastering in our educational institutions. The effective and efficient Web-based teaching/ learning is just getting started and survival is the test for quality assurance.

PUBLISHING ON THE WEB

It is increasingly likely that users may try their hand at setting up their own web pages, and libraries could host these pages. Security will need to be carefully managed as users will need access to the library's server. In addition, there are some legal risks regarding content of such sites and it is advisable to take legal advice. As minimum libraries may need to consider placing a disclaimer on every page making it clear that the library is not responsible for its design and content, and require individuals or groups who publish such pages to display their own name clearly. Having said this however, a useful service which libraries can provide for small firms and organizations is to put up for them, for a small charge, a basic web site of perhaps four pages. The Internet is affecting the twin disciplines of knowledge management, and content management. Knowledge management is the thinking process of converting information to useful knowledge, while the content management is the published information. The author of a Web site must provide the efficient content management, and the visitor who uses a Web site must have the mental ability for an effective knowledge management. The authors need to learn more about the contents alongside the usability of their sites.

ONLINE RESOURCES

The electronic library, providing access to encyclopaedias, dictionaries, newspapers and magazines runs with the help of online resources.

Since the dawn of the Internet age, boosters have predicted the end of leafy college campuses as schools go virtual. The miracle of the Internet was supposed to let great teachers reach any student, any subject, any time, and anywhere.

Online information services are growing too fast to track. Hundreds of organizations of every sort have been putting some basic information sources up on the Web, using sometimes software. And users seem to think they're okay. Community colleges and regional universities that have slowly, organically moved into the online arena — doing their old job in a new way — have succeeded where the flashy business types and big-time private schools have not. Today, the web-based sources offering continues to grow, however, much of the momentum has slowed, and realizing the enormous costs of launching efficacious sources' online. Programs that are technically sound but not fiscally sound may not be endorsed by the library administration because of financial strain to the organization. Conversely, the faculty whom it represents will not endorse information sources that are fiscally sound but not technically sound. The main approach is to develop or maintain services and sources that are technically and fiscally sound.

ONLINE JOURNALS

Online journals in the scientific world, much like their close cousins – online newspapers, are an attempt to harness the powers of computer – mediated communication networks to scientific communication and discourse. The World Wide Web in particular, and the internet and general, have opened new vistas and opportunities to extend the reach of messages via novel and exciting channels and modes of communication. Online journals are already becoming the mode of choice for academic publication. Journal-use studies have traditionally been a way to monitor use of academic libraries' most costly resources. With the advent of online access to full-text journals, many academic libraries find themselves in the position of offering both print and electronic access to journals. In this transitional environment, measuring journal use by both access methods is essential to collection development decisions. While online journals may not soon displace printed journals, they are already a very powerful manner of augmenting traditional forms of exchange. The pre-internet toolbox of communication tools in academe included mostly newsletters, quarterly journals, annual volumes, books and face-to-face seminars and conferences. The online, internet-based journal opens rich and variegated new vistas: these include the ability to search the content, create "Communities of practice" discussion forum where opinions are actually exchanged and explored (rather than simply declared and amplified), create mutual, joint work objects (like common bibliographies), etc. Distance, time, and cost are the first issues. Clearly, online journals allow easy bridging of distances. The internet leaps over continents. Online journals are fast. There is no mail delivery lag, and the cycle of publication is therefore a lot shorter. Online journals are a much less expensive option from the point of view of printing and dissemination. They have no paper or ink costs, minimal connectivity costs, and no mail and delivery costs. Several factors including cost, convenience, and time considerations enter into users' decisions to use online journals. Library users become able to print articles retrieved online at the time of accessing. On the other hand, users had to absorb the costs of photocopying articles from print journals. The hidden printing costs to the library, or to the users printing from their homes or offices, are not as obvious to users as photocopy costs. Printing the online version of an article saves the steps of retrieving and photocopying the original print version, offering obvious time-saving and convenience factors as well. For users accessing the online journals, perhaps the biggest advantage is the elimination of the need to travel to the library. Convenience is one of the most important variables influencing which online journals users select for information accessing.

ONLINE DATABASES

A *database* is structured collection of data. Thus, card indices, printed catalogues of archaeological artefacts and telephone directories are all examples of databases. Databases *may* be stored on a computer and examined using a program. These programs are often called 'databases', but more strictly are *database management systems* (DMS). Just as a card index or catalogue has to be constructed carefully in order to be useful, so must a database on a computer. Similarly, just as there are many ways that a printed catalogue can be organised, there are many ways, or models, by which a computerised database may be organised. One of the most common and powerful models is the 'relational' model and programs which use this model are known as *relational database management systems* (RDMS).

Computer-based databases are usually organised into one or more *tables*. A table stores data in a format similar to a published table and consists of a series of rows and columns. To carry the analogy further, just as a published table will have a title at the top of each column, so each column in a database table will have a name, often called a *field name.* The term *field* is often used instead of column. Each row in a table will represent one example of the type of object about which data has been collected. A database allows to manage and use an incredible variety of information easily. Databases are easy to set-up, easy to manipulate and easy to use. A database allows to maintain order in what could be a very chaotic environment.

ONLINE BOOKS

The online book is an ebook (ELECTRONIC book - a file containing all the information that a hard book would contain). In the olden days, ebooks were sometimes just a plain text file. This is no longer considered industry standard and now, ebooks are carefully formatted like hard books, or they may contain special aspects user can't find in a hard book such as numerous illustrations or even video clips, animations or sound. Many ebooks are available in Adobe PDF format. These can be read on Windows or Mac computers. E-books are in general, not meant for printing out in their entirety but to read on screen or to print out certain sections that are needed when user is off line. Many people do still print them out and bind them, however. Most ebooks are print optimised and will print out easily; some e-books are clearly designed for viewing on computer with many colours, pictures and other features such as in-text hyperlinks and click and point indices.

LIBRARY WEB SITE

Creating a library web site opens up endless possibilities for providing resource based and interactive services such as local information, what's on, book reviews, online magazines, quizzes, online surveys, local history materials, booklists, discographies, and so on. Twenty-four hour access to a virtual library can be made possible. Publishing on the web is getting easier with software which can convert text into html (hypertext markup language) automatically. However, content is not the only consideration. The site needs to be attractive, user-friendly, and easy to navigate. Online tutorials and help pages may be needed, especially for times when staff is not on hand. IT units in local authorities may have experts in web design, or the expertise could be bought in. However, there is an opportunity here for the library to take the lead in designing a corporate web site for the whole authority.

RESOURCE SHARING

The world of electronic information is expanding dramatically. Keeping up with what's new and negotiating advantageous pricing would be practically impossible for any single library. Fortunately, the independent regional networks does this work for users. Based on the specific needs of its members, each network tracks new providers and new offerings from current providers. Using the collective purchasing power of its members, the network offers the library substantial savings on the e-resources users need. Offerings vary by network and include e-books, e-journals and hundreds of full text and citation databases.

ADDITIONAL SERVICES

Due to advances in technology the range of available services is constantly increasing, and with it the expectations of library users. Depending on the aims and purpose of the library, some of the following examples might be regarded as core services, others as value added services:

* OPACs giving in-library access to the catalogue
* remote access to library catalogues
* community information
* seamless access to a range of information sources

* creation of digitised content to provide remote access to special collections and local studies materials
* personalised services to users
* online enquiries services
* requests and renewals by e-mail or other form of remote access
* managed databases of frequently asked questions
* co-operative networked services between libraries
* document delivery
* networking of CD-ROMs
* hosting interactive services for local and national government
* development of services which will extend access to all

IMPACT ON LIBRARY HOUSEKEEPING FUNCTIONS

Perhaps no other recent innovation has impacted the library profession to such a great extent as Internet. Not only is our world becoming an interconnected global community, but this early use of the Internet has changed the fundamental roles, paradigms, and organizational culture of libraries and librarians as well, which created profound impact on L&IS by offering new modes of information delivery and a vast information source. There is a continuing evolution of the roles and functions of libraries and librarians, which appears to parallel the growth of acceptance and use of the Internet by library professionals. The innovative use of Internet technologies enable us to reach both local and distant users much more easily and effectively than hither to possible. Technologies such as email and Web provide tremendous opportunities for Library and Inormation Scientists to deliver the information to the desktops of users. Web offers significant advantage by integrating different library and information services with a common user interface offered by Web browsers. Realizing the potentials, many libraries are rushing to getting the connectivity.

The following description gives an idea of which various functions of libraries may take advantage from Internet and Web technologies.

1. *Acquisition:*
   - Correspondance with Publisher and Vendor.
   - Reminders and Price verification
   - Biblographic details and downloading of bibliographic records etc
   - Ordering, billing
   - Bookshops are on-line e.g. amazon.com
2. *Classification:*
   - Network resources
   - available on the net
   - subscribed or free or trial basis
   - Dewey Online
   - Maths. Classification System
   - Engineering Electronics Library Classification
   - Search engines – such as yahoo use DDC.
3. *Collection Development :*
   - Ownership vs Access
   - Subscribe in print or e-form
   - Subscribe in print as well as in e-form
   - Pay-per-use
   - Consortial approach
4. *Cataloguing:*
   - Cataloguing of network resources
   - Online Catalogues
   - WorldCat (OCLC)
   - WebOPAC – web sites
   - MARC adds 856 field

- OCLC Scorpian project- MARC & AACR2
- Metadata standards- Dublin core

5. *Circulation:*
   - Remote login
   - Status check
   - OPAC access
   - Reminder to users
   - User requests
   - Direct borrowing
   - ILL
6. *Resource Sharing:*
   - Union Catalogue
   - Access, adding, downloading
   - Access to databases over networks
   - Ohionet, ILLINET, WLN, OCLC, BID (UK)
   - full text journals access etc.

Services

- ILL
- Document Delivery Service e.g. Ariel
- Reference / Inf. Services
- CAS
- Recent additions,
- Contents pages
- SDI
- From library collection (Library Catalogues)
- Databases
- Internet Sources
- OPAC
- Database access

- Bibliographical
- Full text
- Many vendors & organisations are moving to Internet (web) access

7. *Subject Lists/ Gateways (With their own Search engines)*
   - Internet Public Library (University of Michigan)
   - EEVL – Engineering
   - SOSIG – Social Science
   - OMNI- Medical
   - ADAM – Arts, Design etc.
8. *User Education:*
   - Through Email
   - Through Web
   - Setting Intranet
9. *Preservation & Storage:*

The Internet is also a medium for the preservation and storage of information. In past, libraries were seen as the main storage facility of information. As society becomes increasingly digital and more information resides on the Internet, the focus on storage and preservation is shifting. For example, some academic libraries are now faced with the problem of whether or not to purchase serials that can be just as easily accessed on – line. Preservation of these same media also becomes an issue of economics, not the "just in case" preservation ideology of the past paradigm. Co joined to the function preservation is the destruction of information. Because the Internet can be seen as a medium for preserving information, the process of destruction of information also is affected. As more and more information is created and stored on the Internet, the capacity to store this information is also decreasing.

HANDLING THE SERVICES

The Internet station(s) are solely designed to access the resources known as the World Wide Web and Telnet. The library may or may not provide access to e-mail accounts, newsgroups or other online forums not falling under the domain of the World Wide Web or Telnet.

The systems must be used for educational, informational and recreational purposes only; not for unauthorised, illegal or unethical purposes. Users must make only authorised copies of copyrighted or licensed software or data. Sometimes it is not possible for library staff to control information contained on specific World Wide Web or Telnet sites. The Library cannot be held responsible for the accuracy of information taken from the public access Internet stations. Usually library staff attempts to assist users in the proper use of any World Wide Web the library uses, but not all staff members will be able to give complete assistance. Since both the Internet and software used to access the Internet rapidly change, the staff may not always be able to give assistance to users

Public Libraries on Telnet Servers

* *Abilene Public Library*
* *Ada Community Library* (login: library) (Boise, ID)
* *Allegan Public Library* (WESTNET) (MI)
* *Allen County Public Library On-Line Catalog* (Ft. Wayne, IN)
* *Anoka County Public Library* (MN)
* *Arlington Heights Memorial Library* (Arlington Heights, llinois)
* *Atlanta/Fulton County Public Library* (Atlanta, GA)
* *Avon Free Public Library* (CONNECT) (CT)
* *Augusta/Mckay Library* (WESTNET) (MI)
* *Augusta-Ross Township Library* (WESTNET) (MI)
* *Baltimore County Public Library* (Towson, MD) enter PAC
* *Bangor Public Library*
* *Beaumont Public Library*

* *Bemis Public Library*
* *Berkeley Public Library* (Berkeley, CA)
* *Berlin-Peck Memorial Library* (CONNECT) (CT)
* *Berrien Springs Community Library* (WESTNET) (MI)
* *Bibliotheque Publique et Universitaire de Neuchatel* (Switzerland)
* *Birmingham Public Library and Board of Education* (Birmingham, AL)
* *Bloomfield Public Library* (CONNECT) (CT)
* *Boonton-Holmes Public Library* (M.A.I.N.) (NJ)
* *Metro Boston Library Network Online Catalog* (includes Boston, Brookline, Cambridge, Chelsea, Lexington, Malden, Newton Public Libraries)
* *Minuteman Library Network* (Boston area) (MLN includes: Acton, Arlington, Bedford, Belmont, Concord, Dedham, ramingham, Lincoln. Medford, Medway, Medfield, Natick, Needham, Norwood, Somerville, Sudbury, Wayland, Waltham, Watertown, Weston, Wellesley, Westwood, Winchester Public Libraries)
* *North of Boston Library Exchange* (Public Access Catalog of the NOBLE Public Libraries includes: Beverly, Beverly Farms, Danvers, Everett, Gloucester, Lynn, Lynnfield, Marblehead, Melrose, Peabody, Peabody—South, Peabody—West, Reading, Revere, Salem, Saugus, Stoneham, Swampscott, Wakefield, Winthrop Public Libraries)
* *Merrimack Valley Library Consortium (MVLC)* (Boston area—MVLC includes: Amesbury, Andover, Billerica, Boxford, Burlington, Carlisle, Chelmsford, Dracut, Hamilton, Haverhill, Ipswich, Lawrence, Lowell, Manchester, Methuen, Newbury, Newburyport, North Andover, North Reading, Rockport, Tewksbury, Topsfield, Wenham, West Newbury Westford, Wilmington Public Libraries)

* *Boulder, Colorado, Public Library System*
* *Bridgman Public Library* (WESTNET) (MI)
* *Bristol Public Library* (CONNECT)
* *Buchanan Public Library* (WESTNET) (MI)
* *Bud Werner / Steamboat Public Library* (MARMOT) (CO)
* *Bur Oak Library System* (ILLINET) (IL)
* *Burlington Public Library* (CONNECT) (CT)
* *Butler Public Library* (M.A.I.N.) (NJ)
* *Canton Public Library* (CONNECT) (CT)
* *Canton Public Library* (Canton, OH)
* *Carroll County Public Library* (MD)
* *Canterbury Public Library* (Christchurch, New Zealand) (logon as opac)
* *Carnegie Library of Pittsburgh's Online Catalog (CAROLINE)* includes menu access to other libraries in the Pittsburgh area
* *Carver County Public Library* (username: SEARCH) (Chaska, MN)
* *Library of the Chathams* (M.A.I.N.) (NJ)
* *Cedar Rapids Public Library*
* *Central and Western Massacusettes Regional System* (C/W MARS)
* *Public Library of Charlotte & Mecklenburg County* (Charlotte, North Carolina) "1995 Library of the Year"
* *Library of the Chesters* (M.A.I.N.) (NJ)
* *Chicago Public Library* (Chicago, IL)
* *Cleveland Public Library*
* *Climax / Lawrence Memorial Library* (WESTNET) (MI)
* *Coloma Public Library* (WESTNET) (MI)
* *Comstock Township Library* (WESTNET) (MI)
* *Corn Belt Library System* (ILLINET) (IL)
* *Cortez Public Library* (MARMOT) (CO)

* *Cumberland Trls Library Syst.* (ILLINET) (IL)
* *Dakota County Public Library* (MN)
* *Delaware Public Library Gopher*
* *Denville Public Library* (M.A.I.N.) (NJ)
* *Denver Public Library*
* *Detroit Public Library*
* *Douglas Public Library District*
* *Dover Public Library* (M.A.I.N.) (NJ)
* *Dowagiac Public Library* (WESTNET) (MI)
* *DuPage Library System* (ILLINET) (IL)
* *Durango Public Library* (MARMOT) (CO)
* *Eagle County Public Library* (MARMOT) (CO)
* *East Hanover Public Library* (M.A.I.N.) (NJ)
* *East Hartford Public Library* (CONNECT) (CT)
* *East Windsor Warehse Point Library* (CONNECT) (CT)
* *Eau Claire District Library* (WESTNET) (MI)
* *Enfield Public Library* (CONNECT) (CT)
* *Enoch Pratt Free Library (PrattCat)* (Baltimore, MD)
* *Estes Park Public Library*
* *Evanston Public Library* (Evanston) (IL)
* *Farmington Library* (CONNECT) (CT)
* *Florham Park Public Library* (M.A.I.N.) (NJ)
* *Fort Collins Public Library*
* *Fort Morgan Public Library*
* *Fort Vancouver Regional Library*
* *Galesburg Memorial Library* (WESTNET) (MI)
* *Galien Township Public Library* (WESTNET) (MI)
* *Garfield County Public Library* (MARMOT) (CO)
* *Glastonbury Memorial Library* (CONNECT) (CT)

- *Glennview Public Library* (Glennview, IL) (login: library)
- *Granby Public Library* (CONNECT) (CT)
- *Grand Rapids Public Library* (Grand Rapids, MI)
- *Great River Library System* (ILLINET) (IL)
- *Greene County Public Library*
- *Harford County Library* (Belair, MD)
- *Hartford Public Library* (CONNECT) (CT)
- *Hartford Public Library* (WESTNET) (MI)
- *Harris County Public Library*
- *Hawaii State Public Library System*
- *Hays Public Library*
- *Hennepin County Public Library* (MN)
- *Hibernia Public Library* (branch of Rockaway Twp. Public Library (M.A.I.N.) (NJ))
- *Houston Area Library Automation Network (HALAN)* (Houston,TX)
- *Houston Public Library* (HALAN) (Houston,TX)
- *Howard County Library* (MD)
- *Ill. Valley Library System* (ILLINET) (IL)
- *Ilsley Public Library (Middlebury, Vermont)*
- *Iowa City Public LIbrary* (Iowa City, Iowa)
- *Jefferson Township Public Library* (M.A.I.N.) (NJ)
- *Kansas City, KS Public Library* (Kansas City, KS) (login: hello visitor.library)
- *Kansas City Public Library*
- *Kaskaskia Library System* (ILLINET) (IL)
- *Keene Public Library*
- *King County Library System (Washington)*
- *Kinnelon Public Library* (M.A.I.N.) (NJ)
- *Lake Hiawatha Public Library* (Branch of Par-Troy Public

Library (M.A.I.N.) (NJ)

* *Lewis & Clark Library System* (ILLINET) (IL)
* *Lawton Public Library* (WESTNET) (MI)
* *Lincoln Park Public Library* (M.A.I.N.) (NJ)
* *Lincoln Township Public Library* (WESTNET) (MI)
* *Lincoln Trail Libraries Syst.* (ILLINET) (IL)
* *Long Hill Township Public Library* (formerly Passaic Township) (M.A.I.N.) (NJ)
* *Lynchburg Public Library*
* *Madison Public Library* (M.A.I.N.) (NJ)
* *Manchester Public Library* (CONNECT) (CT)
* *Maricopa County Library District*
* *Mendham Boro Public Library* (M.A.I.N.) (NJ)
* *Mendham Township Public Library* (M.A.I.N.) (NJ)
* *Marcellus/Wood Memorial Library* (WESTNET) (MI)
* *Marlborough Memorial Library* (CONNECT) (CT)
* *Mesa County Public Library* (MARMOT) (CO)
* *Minneapolis Public Library* (MN)
* *Montville Public Library* (M.A.I.N.) (NJ)
* *Morris County Library* (M.A.I.N.) (NJ)
* *Morris Plains Public Library* (M.A.I.N.) (NJ)
* *Joint Library of Morristown and Morris Township Library* (M.A.I.N.) (NJ)
* *Mountain Lakes Public Library* (M.A.I.N.) (NJ)
* *Mt. Arlington Public Library* (M.A.I.N.) (NJ)
* *Mt. Olive Public Library* (M.A.I.N.) (NJ)
* *Mt. Tabor Public Library* (branch of Par-Troy Public Library (M.A.I.N.) (NJ))
* *Multnomah County Library* (Portland, OR)
* *Nashville Public Library*

* *New Britain Library* (CONNECT) (CT)
* *New Buffalo Public Library* (WESTNET) (MI)
* *New Orleans Public Library* (New Orleans, LA)
* *Newington Library* (CONNECT) (CT)
* *New York Public Library - Branch Libraries Catalog*
* *New York Public Library - Research Libraries*
* *North Suburban Library System* (ILLINET) (IL)
* *North York Public Library* (username: nypl) (North York, ON)
* *Northern Ill. Library System* (ILLINET) (IL)
* *Otsego District Public Library* (WESTNET) (MI)
* *Ottawa Public Library* (no login needed) (Ottawa, ON)
* *Parchment Community Library* (WESTNET) (MI)
* *Parmly-Billings Public Library* (login: library) (Billings, MT)
* *Parsippany-Troy Hills Public Library* (main branch) (M.A.I.N.) (NJ)
* *Pasadena Public Library*
* *Pathfinder System* (MARMOT) (CO)
* *Paw Paw District Library* (WESTNET) (MI)
* *Pequannock Public Library* (M.A.I.N.) (NJ)
* *Pikes Peak Library District* (Colorado Springs, CO)
* *Pitkin County Public Library* (MARMOT) (CO)
* *Plainwell/Charles A. Ransom Public Library* (WESTNET) (MI)
* *Port Arthur Public Library*
* *Port Neches Public Library*
* *Portland Public Library* (CONNECT) (CT)
* *Prince George's County Public Library*
* *Providence (Rhode Island) Public Library*
* *Provo City Library* Catalog (Provo, Utah)

* *Pueblo Library District* (Pueblo, Colorado)
* *Queens Borough Public Library InfoLinq* (login: QUEENS)
* *Ramsay County Public Library*(login: library) (Shoreview, MN)
* *Randolph Public Library* (M.A.I.N.) (NJ)
* *Regina Public Library* (login: public) (Regina, SK)
* *Richland Community Library* (WESTNET) (MI)
* *River Bend Library System* (ILLINET) (IL)
* *Rockaway Boro Public Library* (M.A.I.N.) (NJ)
* *Rockaway Township Public Library* (M.A.I.N.) (NJ)
* *Rocky Hill Library* (CONNECT) (CT)
* *Rolling Prairie Library Syst.* (ILLINET) (IL)
* *Roxbury Public Library* (M.A.I.N.) (NJ)
* *Salt Lake City Public Library*
* *San Francisco Public Library* (San Francisco, CA)
* *San Miguel County Public Library* (MARMOT) (CO)
* *Saskatoon Public Library* (username: public; at the PAC >>> prompt, type pac) (Sasksatoon, SK)
* *Scott County Public Library* (MN)
* *Seattle Public Library* (Seattle, WA) — 1st to provide public access to the Internet!
* *Shawnee Library System* (ILLINET) (IL)
* *Simsbury Public Library* (CONNECT) (CT)
* *Sioux City Public Library* (Sioux City, Iowa)
* *Sodus Township Library* (WESTNET) (MI)
* *Sonoma County Library*
* *South Haven Memorial Library* (WESTNET) (MI)
* *South Windsor Public Library* (CONNECT) (CT)
* *Southington Public Library* (CONNECT) (CT)
* *Southwest System* (MARMOT) (CO)

* *Spokane Public Library* (Spokane, Washington)
* *Springfield-Greene County Library* login: public (Springfield, MO)
* *Starved Rock Library System* (ILLINET) (IL)
* *Sterling Municipal Library*
* *Sterling Public Library*
* *St. Joseph County Public Library On-Line Catalog* (South Bend, IN)
* *St. Joseph/Maud Preston Palenske Memorial Library* (WESTNET) (MI)
* *St. Paul Public Library* (MN)
* *Sterling Municipal Library*
* *Sterling Public Library*
* *Suffield Memorial Library* (CONNECT) (CT)
* *Suburban Library System* (ILLINET) (IL)
* *Summit County Public Library* (MARMOT) (CO)
* *Three Oaks Township Library* (WESTNET) (MI)
* *Three Rivers System* (MARMOT) (CO)
* *Vail Public Library* (MARMOT) (CO)
* *Vancouver Public Library* (login: netpac; password: netpac1) (Vancouver, BC)
* *Vicksburg Community Library* (WESTNET) (MI)
* *Washington County Public Library* (MN)
* *Washingtown Township Public Library* (M.A.I.N.) (NJ)
* *Watervliet District Library* (WESTNET) (MI)
* *Western Ill. Library System* (ILLINET) (IL)
* *West Bloomfield Public Library*
* *West Hartford Public Library* (CONNECT) (CT)
* *Westerville Public Library*
* *Wethersfield Public Library* (CONNECT) (CT)

* *Wharton Public Library* (M.A.I.N.) (NJ)
* *Whippanong Library Association* (M.A.I.N.) (NJ)
* *Wilmington Library, Delaware*
* *Windsor Locks Public Library* (CONNECT) (CT)
* *Windsor Public Library* (CONNECT) (CT)

Public Libraries with Gopher Servers

* *Allen County Public Library Gopher* (Ft. Wayne, IN)
* *Boston Public Library Gopher* (Boston, MA)
* *British Columbia Library Services Branch Gopher* (with addresses & telephone numbers of 74 British Columbia public libraries)
* *Chester County Library and District Center* (Chester County, PA)
* *Cleveland Public Library Gopher* (Cleveland, OH)
* *Delaware's Public Library Gopher*
* *Denver Public Library*
* *Queens Borough Public Library Gopher*
* *Salt Lake City Public Library Gopher* (Salt Lake City, UT)
* *Pikes Peak Library's MaggNet*
* *Public Library of Youngstown and Mahoning County* (Youngstown, OH)

REFERENCES

1. Budd, Richard & Ruben, Brent., (ed) (1979). *Interdisciplinary approaches to human communication.*. Rochelle Park, NJ: Hayden, 71-93.
2. Cameron, C., (December 15, 1997). *Theories and metaphors of cyberspace.* Available at: http://pespmc1.vub.ac.be/CYBSPASY/CCameron.html.
3. Canter, David. , Rivers, Rod. & Storrs, Graham. (1985). "Characterizing user navigation through complex structures". *Behavior and Information Technology*, 4(2), 93-102.
4. Canter, David, Powell, J., Wishart, J., Roderick, C., (1986) "User navigation in complex database systems." *Behaviour and Information Technology*, 5(3), 249-257.

5. whatis.techtarget.com/definitionsCategory/0,289915, sid9_tax 1672,00.html
6. *www.wiltonlibrary.org/innovate.html*
7. *www.adobe.com/resources/business/rich_internet_apps*
8. theopensourcery.com/xmlria.htm
9. *www.computerworld.com/taxonomy/000/000/100/taxonomy_000000168_index.jsp*
10. www.digital-web.com/articles/usability_for_ rich_internet_applications
11. Coyne, Richard (1995). *Designing information technology in the postmodern age: from method to metaphor.* Cambridge, MA: MIT Press, 249-301.
12. Cunningham, Michael (December 30, 1996). "*An A-Z of Internet metaphors*". *Irish Times*, Available at: http://www.irish-times.com/irish-time/paper/1996/1230/cmp2.html.
13. Davis, James, (June 11,1997). "Mixed metaphors", alt.humor.puns
14. Deetz, Stanley & Mumby, Dennis, (1985). "Metaphors, information and power." In Ruben, B., ed. *Information and human behavior.* Transactions, 1, New Brunswick, NJ: Rutgers University Press. pp.369-385
15. Gold, David (1997). "*You can't surf a sine wave: metaphors and the future of the Internet.*" Available at: http://ccwf.utexas.edu/~dgold/metaphor.project/.
16. Jamieson, Kathleen Hall, (1980). "The metaphoric cluster in the rhetoric of Pope Paul VI and Edmund G. Brown Jr." *Quarterly Journal of Speech*, 66, 51-72.
17. Johnson, Steven, (October 23, 1997). "Metaphor monopoly." *New York Times*, A 27.
18. Palmquist, Ruth, (1996). "A qualitative study of Internet metaphors", *17th National On-line Meeting*, May 1996.
19. Palmquist, Ruth, 1996, "*The search for an Internet metaphor: a comparison of literatures*", *American Society of Information Science Conference*, October 19, 1996. Available at: http://www.asis.org/annual-96/Electronic Proceedings/ palmquist.html
20. Rohrer, Tim, (1995). "*Metaphors we compute by: bringing magic into interface design*". Available at: http://philosophy.uoregon.edu/metaphor/gui4web.htm
21. Rohrer, Tim (1997) "Conceptual Blending on the Information Highway: How do metaphorical inferences work?" in Discourse and Perspective In Cognitive Linguistics, edited by Wolf-Andreas Liebert, Gisela Redeker and Linda Waugh. Amsterdam: John Benjamins, 1997, pp. 185-205. [Also available at: *http://philosophy.uoregon.edu/metaphor/iclacnf4.htm*]

22. Ross, P.E. (1992). "New whoof in Whorf" *Scientific American*, 266 (2), 24-25. Clarke, Roger, (1996). "*A more refined populist metaphor*". Available at: http://www.anu.edu.au/mail-archives/link/linl9604/0032.html
23. Coblentz, Clayton, (1997). "*Metaphorically interactive: the rhetorical examination of World Wide Web pages*". Available at: http://www.regent.edu/acad/schcom/rojc/coblentz/coblentz.html
24. Lawler, John, "*Metaphors we compute by*", Lecture to the Informational Technology Division of the University of Michigan. Available at: (http://www.lsa.umich.edu/ling/jlawler/meta4compute.html). [Original version of the essay appearing in, Hickey, D.J. (1999) Figures of thought: for college writers. Mountain View, CA: Mayfield Publishing.)
25. MacCormac, Earl (1976) *Metaphor and myth in science and behavior,* Durham, NC: Duke University Press.
26. Norton, Catherine, (1989) *Life Metaphors*, Carbondale, IL: South Illinois University Press.
27. Osborne, Michael, (December 1977). "The evolution of the archetypical sea in rhetoric and poetic", *Quarterly Journal of Speech, 63*(4), 347-363.
28. Ostman, Charles, "*The Internet as organism.*" Available at: http://pespmc1.vub.ac.be/CYBSPASY/COstman.html.
29. Sawhney, Harmeet (1990). "Information superhighways, metaphors as midwives", *Media, Culture and Society, 18(*29),314.
30. Seligman, Doree Duncan and Laporte, Cati & Bugaj, Stephan Vladimir, (1998). "*The message is the medium.*" *Sixth International World Wide Web Conference*, Bell Laboratories, Lucent Technologies. Available at: http://www.6conf.org/HyperNews/get/PAPER119.html.
31. Rothstein, Edward, (October 28, 1996) "Connections: metaphors for Internet may affect its use." New York Times.
32. Smith, Ruth C. & Turner, Paige K. (June, 1995). "A social constructionist reconfiguration of metaphor analysis", *Communication Monographs*, 62, 152-181.
33. Stefik, Mark, (1997). *Internet dreams: archetypes, myths and metaphors.* Cambridge, MA: MIT Press.
34. Thaler, Lee (ed), (1970). *Communication: general semantics perspectives,* New York, NY: Spartan Books.
35. White, Peter, (October 29, 1996). "*On-line services and transaction space: conceptualizing the issues*". Available at: http://dpub36.pub.sbg.ac.at/ectp/WHITE_P.htm
36. Whorf, Benjamin, (1939). Science and Linguistics, *Technology Review,* 42, 247-248.

11
Library Networks

A library network is broadly described as a group of libraries coming together with some agreement of understanding to help each other with a view to satisfying the information needs of their clientele. UNISIST II working document defines Information Network as:

A set of inter-related information systems associated with communication facilities, which are cooperating through more or less formal agreements in order to implement information handling operations to offer better services to the users.

The National Commission on Libraries & Information Science in its National Programme Document (1975) defines a network as:

Two or more libraries engaged in a common pattern of information exchange, through communications for some functional purpose.

OBJECTIVES

* To promote and support adoption of standards in library operations.
* To create databases for projects, specialists and institutions to provide online information services
* To improve the efficiency of housekeeping operations
* To coordinate with other regional, national &

international network for exchange of information and documents

* To generate new services and to improve the efficiency of existing ones

FUNCTIONS OF LIBRARY NETWORKS

1. Digitization and Preservation

Understanding the options for digitization and preservation and selecting the right combination of services can be challenging for even the most experienced. Some networks have developed a specialization in the areas of digitization and preservation and serve as a resource for their colleagues at other networks. Working through the network, user can tap into the expertise of the networks that have invested in building a robust digitization/preservation program. Services are available include consulting on digitization options and priorities, education and training, identification of funding sources, and grant assistance.

2. Professional Training

Rapidly changing technology and demand for new services create training challenges for library staff. Most of the library network design a professional development program to meet the specific needs of its members. Some networks offer online as well as face-to-face courses and customized courses as requested. Courses range from management and technology to hands-on training for general use software and library services. In addition, many networks may participate in the Network Education Exchange in future, a cooperative online training program that maximizes expertise and extends training capabilities by sharing their online courses with all the members of the participating networks.

3. Library Tools

The network can help libraries to achieve cost savings. Based on member needs, each network seeks out desired service providers and negotiates for pricing. The range of products is broad and varies by network. Types of products include anything from barcode labels and other library supplies to hardware and software, serials management, virtual reference, federated search software and much more. In addition to providing a diverse set of library products and services, the library network also serves as a source for library services. As these services have increased in number and complexity, it's good to know that network is there to help.

LIBRARY NETWORK DEVELOPMENT IN INDIA

Some factors that are responsible for the development of library and information networks in India are:

* The report of the working group of the planning commission on modernization of library services and informatics for the seventh five year plan, 1985-90
* The National Policy on Library & Information systems document (1986) accepted by the ministry of HRD, Government of India.
* The report on national policy on university libraries prepared by the Association of Indian Universities (1987)
* The UGC report on information systems for science and technology under the Department of Science & Industrial Research (DSIR) Government of India has been vigorously promoting an integrated approach to library automation and networking.

DELNET (Developing Library Network)

Sponsor: NISSAT & NIC (1988)

Objective: To promote resource sharing; develop a network of libraries; collect, store, disseminate information

Members: 165 Institutions, 600 Libraries, 15 States in India, 5 from outside India

Services: resource sharing; free Software; ICE online facility; books database; thesis database; Indian specialists database

DELNET has been in operation since January 1988 and was registered as a society in 1992. It was initially sponsored by the National Information System for Science and Technology (NISSAT), Department of Scientific and Industrial Reseach, Government of India and is currently being promoted by the National Informatics Centre, Department of Information Technology, Ministry of Communications and Information Technology, Government of India and India International Centre, New Delhi.

DELNET has been established with the prime objective of promoting resource sharing among the libraries through the development of a network of libraries. It aims to collect, store, and disseminate information besides offering computerised services to users, to coordinate efforts for suitable collection development and also to reduce unnecessary duplication wherever possible.

The Main Objectives of DELNET are:

* To promote sharing of resources among the libraries by developing a network of libraries, by collecting, storing and disseminating information and by offering computerised services to the users;
* To undertake scientific research in the area of Information Science and Technology, create new systems in the field, apply the results of research and publish them;
* To offer technical guidance to the member-libraries on collecting, storing, sharing and disseminating information;
* To coordinate efforts for suitable collection development and reduce unnecessary duplication wherever possible;
* To establish /facilitate the establishment of referral and /or research centres, and maintain a central online union catalogue of books, serials and non-book materials of all the participating libraries;
* To faciliate and promote delivery of documents manually or mechanically;

* To develop specialised bibliographic database of books, serials and non-book materials;
* To develop databases of projects, specialists and institutions;
* To possess and maintain electronic and mechanical equipment for speedy communication of information and delivery of electronic mail;
* To coordinate with other regional, national and international networks and libraries for exchange of information and documents;

DELNET Databases

a) Union Catalogue of Books

DELNET maintains an online union catalogue of books available in its member-libraries. This union catalogue is continuously updated and is growing in size. The information can be retrieved by author, title, subject, conference, series, etc. It has 53,95,778 bibliographic records. The request for inter-library loan can be placed through the online system.

b) Union List of Current Periodicals

DELNET has created union lists of current periodicals in science and technology, social sciences and humanities. This database is available online to DELNET users. It now lists 25,252 periodicals and is regularly updated and new titles are added annually. It is a major resource for Document Delivery Services.

c) Union Catalogue of Periodicals

DELNET maintains a union catalogue of periodicals, which contains full holdings data of the libraries. At present, the database contains 20,235 records.

d) Database of Periodical Articles

The database has details of articles which can be searched under the title, author, compiler, name of the periodical and subject. The database is being extensively utilised by the researchers and scholars. At present the database contains 8,88,342 records.

e) CD-ROM Database

A bibliographic database of CD-ROMs available with the member-libraries is being compiled. It has 2,281 records.

f) Union List of Video Recordings

This is a database of video cassettes available in DELNET member-libraries and has about 5,000 listings.

g) Union List of Sound Recordings

This union list consists of audio cassette records available in member-libraries. This database has 748 records.

h) Database of Theses and Dissertations

A database of Theses and Dissertations submitted to Indian Universities has been started, which covers various subjects. The database has 44,304 records.

i) Union List of Newspapers

The database has 70 records and contains information about the newspapers including title, name of the editor, published from, E-mail address and also the Web address of the INTERNET edition if available on the WWW.

j) DEVINSA Database

DEVINSA (Development Information Network for South Asia) is a special database on socio-economic issues. It has nearly 20,000 records of periodical articles, books and unpublished material on socio-economic issues.

k) Profile of Member-Libraries

A directory of member-libraries is available and contains information about them

II Delnet also provides access to

a. Cambridge Dictionaries online

b. Networked Digital Library of Theses and Dissertations

c. GISTNIC Databases
d. MEDLINE & other databases of NLM
e. U.S. Patents: Full Text
f. Full Text Medical Journals
g. Engineering & Technology E-Journals:
h. Table of Contents etc.

Database statistics

Delnet Database	*No. of Records*
Union CataLogue of Books - CCF	53,95,778
Union List of Current Periodicals	25,252
Union Catalogue of Periodicals	20,235
Databases of Periodical Articles	8,88,342
CD-ROM Database	2,281
Union List of Video Recordings	5,000
Union List of Sound Recordings	748
Databases of Theses and Dissertations	44,304
DEVINSA Database	20,000

DELNET Loan and Document Delivery Services

DELNET is offering inter-Library loan & Document Delivery Services to its Member Libraries. ILL requests can be registered online for books. For the resources not available in the union catalogues and journal articles, requests can be sent to DELNET through E-mail. DELNET has also prepared ILL Guidelines for use by the member-libraries. The services are quite popular among the member-libraries

Retro Conversion

DELNET undertakes retro-conversion projects periodically.

Professional Service/Training

DELNET arranges tutorials, workshops, lectures, and training programmes every year from time to time besides a National Convention on Library and Information Networking (NACLIN).

DELNET also offers online information on

Announcements of the International Federation of Library Associations and Institutions (IFLA) and Library Science jobs globally. This information provides inputs to library and information professionals for improving their talent and professional expertise.

National bibliographic database

DELNET undertook the compilation of the National Bibliographic Database with the support of the Department of Culture, Government of India in 1998 as a Pilot Project. Fifty thousand records of books: 20,000 in English, 15,000 in Tamil and 15,000 in Punjabi were created at the Punjabi University, Patiala and International Institute of Tamil Studies, Chennai. At Andhra University Library, Visakhapatnam, 15,000 records in English and 10,000 records in Telugu have also been created. Similarly 25,000 records have been created at the Asiatic Society, Mumbai. The work is in progress at the Asiatic Society, Kolkata.

Software Developemt

(i) DEL-DOS

DEL-DOS, based on DOS platform is used for creating MARC records of books published in English as well as for creating records in Indian languages using GIST technology. Some of its salient features are :

1. The software is simple and easy to use. Any novice library professional or the end user can use the software without any prior training.
2. It is user-friendly for data inputting and search purposes.
3. The records can be stored in MARC format.
4. Deletion and printing of selective and range of records is possible.
5. Global values can be defined during the worksheet entry.
6. The search results can be displayed in the Tag format, AACR-II or the brief summary can also be displayed and stored or printed out.

7. The name authority database fe maintained.
8. Editing of records is very simple and fast.
9. Import/Export into and from ISO 2709 is possible.

(ii) DEL-MARC

1. Software designed and developed exclusively to work in all kinds of libraries including public libraries, academic libraries and special libraries.
2. Multi-user facility of this software will allow different branch libraries/sections/nodes to function together.
3. Free upgradation.
4. Follows internationally recommended standards and formats such as US-MARC, MARC 21
5. Available at a very low cost.
6. Barcode enabled.

DELMARC manages libraries with extensive collections, complex operations and serving a large clientele. This software has been designed after a comprehensive study of different library-related functions performed in different kinds of libraries. This software is easy to work with and offers the following modules at this stage:

* Acquisition
* Cataloguing
* Circulation
* OPAC
* Administration
* Export/Import
* Report Generation
* Barcoded stock verification
* Article Indexing
* Serial control being tested
* Import Facility from external files
* Authority Control

(iii) DEL-PLUS

1. Software designed and developed exclusively to work under all kinds of libraries.
2. Free upgradation
3. Follows internationally recommended standards and formats such as MARC 21
4. Best for small and medium size libraries which have collections upto one lakh holdings.
5. Barcode enabled.

This software is easy to work with and comprises the following modules:

* Acquisition
* Cataloguing
* Circulation
* OPAC
* Administration
* Export/Import
* Authority Control
* Article Indexing
* Report Generation
* Stock Verification

The modules for Serials Control are being tested and will be released soon.

(iv) DEL-VISION (A multilingual retro conversion Application)

In order to modernise and network the Libraries in India with regional language facility, DELNET presents Del-VISION (UNICODE based, Multilingual), Retro-conversion software pacdkage, conforms to International standards and presents its application in user-friendly manner.

DELNET has now released the first version of Del-Vision, multi user LAN based retro-conversion software. Del-Vision has an effective facility to work with multiple languages at the same time using MARC formats.

Technical Support

DELNET adopts the latest dependable technologies in information science. This helps libraries to get advice from DELNET to adopt technologies that emerge periodically. This saves expenses and time of the member-institutions.

DELNET offers technical support to member institutions in the selection of:

a. Hardware
b. Software
c. Communication Links
d. Database Creation; and
e. Technical problems faced by member-libraries from time to time.

General Networks in India:

NICNET

Title : National Information Center Network

Sponsor : Planning Commission, Govt. of India.

Membership : Four national and regional nodes, 32 state and union territory nodes; seventy cities and towns

Services : Bulk file transfer; teleconferencing; full text and bibliographic retrieval services

Application : ICMRNIC Center; MEDLARS in India; Chemical Abstracts database

National Informatics Center is the nodal Information Technology organization for informatics development and networking in Government and Government related Organizations . NICNET facility has been established in all Central Government departments, 35 States/UTs and about 600 District centers to facilitate informatics development for decision support and information exchange. A high speed NICNET National info-Highway, an incremental overlay over the existing network had been setup with SCPC , FTDMA, TDMA VSATs, DVB Broadband VSAT technology and

Wireless MAN to take advantage of Internet technology in India.

National Informatics Centre (NIC) of the Department of Information Technology is providing network backbone and e-Governance support to Central Government, State Governments, UT Administrations, Districts and other Government bodies. It offers a wide range of ICT services including Nationwide Communication Network for decentralised planning, improvement in Government services and wider transparency of national and local Governments. NIC assists in implementing Information Technology Projects, in close collaboration with Central and State Governments, in the areas of (a) Centrally sponsored schemes and Central sector schemes, (b) State sector and State sponsored projects, and (c) District Administration sponsored projects. NIC endeavours to ensure that the latest technology in all areas of IT is available to its users.

The organizational set up of NIC encompasses its Headquarters at New Delhi, State Units in all the 28 State capitals and 7 Union Territory Headquarters and District centres in almost all the Districts of India. The Organization employs a large pool of efficient technical manpower.

At the NIC Headquarters, a large number of Application Divisions exist which provide total Informatics Support to the Ministries and Departments of the Central Government. NIC computer cells are located in almost all the Ministry Bhawans of the Central Government and Apex Offices including the Prime Minister's Office, the Rashtrapati Bhawan and the Parliament House.

Nationwide projects

NIC has conceptualised, developed and implemented a very large number of project for various Central and State Government Ministries, Departments and Organisations. Many of these projects are continuing projects being carried out by various divisions of NIC at New Delhi Headquarters and State/

District centres throughout the country. Some of the most noteworthy projects are presented to offer the viewers/citizens a glimpse of the multifaceted, diverse activities of NIC, touching upon all spheres of e-governance and thereby influencing the lives of millions of citizens of India.

Services

* Corporate Communications
* Internet/Intranet
* E-mail
* EDI
* Web enabled database access
* Emergency voice communication
* Data broadcast
* Multimedia broadcast
* Multicasting
* Distance Learning

INDONET

Title : INDONET data Network

Sponsor : CMC Ltd (1986) = Informatics India Ltd (1989)

Membership : Commercial computer network

Services : Database services such as DIALOG, COMPUSERVE; IP; SHARP

Applications : ACME; file transfer; international gateway

CMC's Indonet is India's first Value Added Services & Data Network.

It is approved by the Department of Telecommunications and is TCP/IP based.

INDONET has:

* High speed leased links connecting to 8 locations in India.
* At least 2 RISC based servers at each location.
* ISDN backup at each location, for better reliability and up-time.

* Multiple high speed Internet gateways.

It offers the following services as of today:

E-Mail / Web-based solutions

It offers Internet Mail/Fax/File transfer/EDI services. It also develop corporate Intranet/Web-based applications...

Database management and Data Processing services

It provides Database services/Job processing/Computerization/Hosted tailor made applications/Remote computing etc...

Interactive Voice Response Services

I - NET (VIKRAM)

Title : I - NET

Sponsor : Dept. of Telecommunications, Govt. of India

Connectivity : Packet switched public data network covering nine cities

Services : Information exchange through e-mail / FTP; Bibliographic databases

CALIBNET

Title : Calcutta Libraries Network

Sponsor : NISSAT - Govt. of India

Applications : Cataloging; serials control; acquisitions; circulation

Services : CAS; SDI; union catalog; partial database; editing and retrieval of records; global information; search; full-text document delivery; library automation; CALIBNET INFO Services

CALIBNET, a Government of India Project, has been launched by the National Information Systems for Science and Technology (NISSAT), Department of Scientific & Industrial Research (DSIR); and managed by the CALIBNET Society established under the West Bengal Government's Societies Registration Act 1961. CALIBNET aims to provide the individual libraries and their reading members with cost-effective solutions to their information problems. The primary objective of the Project is building access to library & information resources available in the eastern region. This has been pursued through :

Implementation of a series of Databases, Bibliographic Factuals. and Intellectual Assets of West Bengal this apart, CALIBNET provides its Participating Members and the user community at large with the following :

* Indian Library & Network Resources
* Overseas Library Resources On India
* Worldwide Library Catalogues
* National Libraries of the World
* Newspapers & Journals
* Electronic Reference Tools
* Factual Information Sources
* Document Supply Services, and

Varied Programs for

* On-Demand Information Services
* Consultative Service for Library Automation
* Manpower Development Opportunities, and
* Research & Development in IT Applications

BONET

Title : Bombay Library Network

Sponsor : NISSAT & NCST (1994)

Objective : To promote cooperation among libraries in Bombay

Services : online catalog; online document delivery; IRS; interlibrary loan; dissemination of information

ADINET

Title : Ahmedabad Library Network

Sponsor : NISSAT, DSIR (1994) & INFLIBNET

Objective : To bring cooperation among its regional

libraries; to develop databases; to integrate scientific and technical information systems

Members : nine libraries

Services : library automation; library holdings; database in progress

LIBRARIES are rightly called the "store houses" of valuable knowledge, in the form of invaluable books, journals and other information material. With the rapidly growing advancement in every field, more and more documents are becoming available the world over in printed and machine readable forms. No library can afford to house every necessary document in the area of its interest. Resource sharing, thus assumes great importance at this juncture and such a solution is being effectively provided by the "Ahmedabad Library Network" (ADINET) to the users and librarians specializing in any discipline of knowledge and stationed anywhere in the world.

ADINET is an Information Network of Libraries in and around Ahmedabad. ADINET was registered as a Society in October 1994. It was sponsored by National Information System for Science and Technology (NISSAT), Department of Scientific and Industrial Research, Government of India.

Objectives

* To bring about cooperative mode of working amongst libraries and information centers in and around Ahmedabad.
* To integrate the economic, scientific and technical information systems into an effective network in and around Ahmedabad.
* To facilitate and promote sharing of resources amongst the libraries and information centres in and around Ahmedabad by developing and maintaining a central on-line Union Catalogue containing bibliographic

information on books, serials and non-book materials of all the participating libraries.

* To coordinate with other regional, national and international networks, libraries, information and documentation centres for exchange of information and documents.
* To offer technical guidance to member libraries on selecting, storing, sharing and disseminating information.
* To coordinate efforts for suitable collection development and reduce unnecessary duplication whenever possible.
* To develop databases of projects, specialists and institutions in and around Ahmedabad.
* To create awareness amongst all users' groups and to educate them in the utilisation of information.
* To develop resources and to propagate information in ways appropriate to the needs of users in and around Ahmedabad.
* To help library and information centre users and also individuals who practice different professions in getting specialized information of their interest

Services Offered By ADINET

(a) Data base of current periodicals

ADINET is maintaining database of current periodicals received by over 100 libraries in and around Ahmedabad. Paper copy, floppy & on-line versions are available.

This is extensively useful to librarians of various types of libraries.

(b) Inter-Library Loan

ADINET enables a user to find location of periodical/book/report of his/her interest. Books, Reports, Theses and even non-book material such as audio-video material can then be borrowed on inter-library loan basis.

Inter-Library document delivery service is available.

(c) CUrrent COntents for Library and Information Science [CUCOLIS]

Content pages of Library & Information Science journals are regularly supplied to all members. Full texts of selected articles from content pages are supplied on request.

(d) Supply of Photocopies from Journals received by libraries in and around Ahmedabad

Supply of photocopies from journals received by libraries in and around Ahmedabad is supplied on request.

(e) Database of Databases available in Lib. & Inf. Centres in and around Ahmedabad

ADINET is preparing the database of databases available in library and information centres in and around Ahmedabad. ADINET acts as a pointer towards various databases/ information available with the institutions/professional associations.

(f) Job opportunities for librarians (JOLI)

ADINET is maintaining the database known as JOLI of fresh qualified Library and Information Science students and other qualified library professionals who want to change the job for better prospects. This is extensively used as ADINET is consulted by many institutions for their requirements of library trainees as well as for good well trained/ experienced Library and Information professional for permanent post of librarians.

(g) Special Internet training to fresh qualified librarians

ADINET imparts such training to fresh qualified librarians. They are specially trained for Internet surfing for particular subject/theme. ADINET helps them to get suitable job.

(h) ADINET STUDY CIRCLE AND MEETINGS

ADINET Study Circle Meetings are held to discuss professional problems, developments, etc and share the experiences.

(i) Training Programmes

ADINET is giving computer training to Library Professionals by Organising Training programmes on CDS/ ISIS and WINISIS regularly. Training to individual librarian is also imparted. Various training programmes are also conducted.

(j) Ranganathan's Day celebration

ADINET every year celebrates the birth anniversary of Dr. S. R. Ranganathan (Father of Indian Library and Information Science) as Librarians' Day. It is formally inaugurated in the morning and followed by a seminar on current topic of interest to people connected with library and information centres.

(k) New Services introduced / offered recently

It is the present observation that most of the libraries are unable to update their day to day library jobs due to budget crunch on one hand and lack of professional staff on the other. They need some professional/organization who can assist in completing their backlog and also help them continuously for betterment of their libraries without any regular & permanent financial commitments.

One of the objectives of ADINET is to bring about cooperative mode of working amongst the libraries and information centers in and around Ahmedabad. In order to foster mutual cooperation and assistance among the libraries for the development of library services, ADINET is now offering the following services by the special expert team against reasonable charges. ADINET wants to help libraries and information centers at their shop floor levels.

At present following services are available to libraries and information centers in and around Ahmedabad & Gandhinagar.

(1) Computerization of library.

(2) Cataloguing, classification of library documents either manually or by computer operation.

(3) Labeling and Shelving of books.

(4) Stock verification of library documents.
(5) Manpower development programmes.
(6) Planning for library development.
(7) Completion of backlog of any work of library and information centers.

(m) Jobs For Librarians

(n) Free Listing of qualified Librarians on ADINET website

All qualified librarians working in libraries of Gujarat are invited to send their name, address, qualifications, designation, library's name, address, telephone numbers, e-mail address. Only condition for listing is that the person must have his/her e-mail address. Send required information only through e-mail to ADINET (alibnet@gmail.com)

Publications

* CUrrent COntents for Library and Information Science [CUCOLIS]
* Directory Of Institutions And Colleges In Ahmedabad & Gandhinagar And University In Gujarat - 2007
* Union List of Current Journals. Every year ADINET brings out an Union list of current journals subscribed by libraries in Ahmedabad and Gandhinagar. This information has been found quite handy and useful by the libraries and their users.

MYLIBNET

Title : Mysore Library Network

Sponsor : NISSAT (1994)

Objective : Developing software tools; conducting seminar; workshops/training programs; conduct surveys

Host Site : CFTRI, Mysore

Members : 116 Institutions

Services : MYLIB Database; E-journals; food patents; CFTRI Library Bulletin; public services.

The National Information System for Science and Technology (NISSAT), Dept. of Scientific and Industrial Research (DSIR) took up initiatives for the design and development of library networks in India in 1985 in order to share the resources available in the libraries, located in various parts of the country. The Mysore Library Network was setup during May 1995 in the city of Mysore under the financial assistance from NISSAT. The Mysore Library Network is housed inside Central Food Technological Research Institute (CFTRI) campus.

There are 116 colleges/institutions affiliated to University of Mysore, out of which 34 colleges are located with in Mysore city. All the libraries within Mysore city will be networked in the first phase with the following objectives :

* To share resources available with all the libraries.
* To provide a faster communication to all the libraries through Electronic Mail facility. To develop software tools for better library management.
* To create awareness in the field of latest Information Technology by conducting seminars/workshops/training programmes.
* To setup a Information base in collaboration with industries.
* To conduct surveys.
* To flash arrival of new books/journals, announcement of events like seminar/workshop/training programmes.

COUNTRYWIDE AREA NETWORK

DESINET

Title: Defence Science Information Network

Sponsor: DESIDOC, Delhi

Activity: Focus on scientific, research and defense communities

ERNET

Title : Educational and Research Network

Sponsor : Dept. of Electronics, Govt. of India; UNESCO (Financial assistance from UNDP)

Members : eight institutions (5 IITs, IISc., National Centre for Software Technology - Bombay, CCI wing of Dept. of Electronics)

Services : Communication services such as e-mail, file transfer, remote log on, database access, bulletin board etc.,

SIRNET

Title : Scientific and Industrial Research Network

Sponsor : CSIR (Commissioned Agency- NCST, Bombay)

Members : 40 labs and R&D Institutions

Applications : scientific communication; leather technology; natural products; food technology; medicinal Plants

VIDYANET

Title : VIDYANET (Dedicated Communication Computer Net)

Sponsor : TATA Institute of Fundamental Research, Bombay

Objectives : To provide rapid means of communications by linking computers at various institutions in India to similar networks outside the country; to stimulate corporate research, the day-to-day exchange of research information and the execution of joint projects and publications

Services : File transfer facility; sharing of computer resources and access to remote applications, databases, libraries, etc.

BTISNET

Title : BTISNET (Specialized Information Network)

Computer Science Journals	*Indian Medical Plants*
http://www.lib.nus.edu.sg/slb/ejcomp.html	http://www.indmedplants-kr.org/
informationr.net/ir/titlepages.html	
www.emis.de/journals	
http://www.waikato.ac.nz/library/learning/g_services.shtml	
http://www.shef.ac.uk/library/subjects/subcmsci.html	
www.math.vt.edu/people/linnell/Electronic/	
http://www.kcl.ac.uk/depsta/iss/schools/hlsp/pse/engint.html	
library.uthct.edu/interf012.htm	
http://www.staffs.ac.uk/services/library_and_info/engineering/journ.htm	
http://www2.lib.ku.edu/~watsonreference/edocs.htm	
library.ukc.ac.uk/library/info/subjectg/healthinfo/Ejournals.htm	
http://www.staffs.ac.uk/services/library_and_info/compjnls.htm	
http://www.nitro-net.com/internet.htm	
http://devweb.library.unr.edu/ejournals/subject/cos.html	

Sponsor : Dept. of Biotechnology, Govt. of India.

Connectivity : 10 Specialized Information Centres in genetic engineering, plant tissue culture; photosynthesis and plant molecular biology; cell transformation ; bio-process engineering.

Services : Data processing using applications software; online communication access; facsimile facility

INFLIBNET

Title : Information Library Network

Sponsor : UGC (1991)

Connectivity : computer communication network of universities and R&D; libraries and bibliographic information centers throughout the country

Members : 200 Universities; 400 College libraries; 200 R&D libraries

Services : catalog service; database Services; document supply services; e-mail; BBS: audio and video conferencing, etc.

BALNET

Title : Bangalore Library Network

Sponsor : JRD;. Tata Memorial Library (1995)

Members : 100 Libraries

MALIBNET

Title : Madras Library Network

Sponsor : INSDOC & NISSAT (1993)

Members : 15 Libraries

Activity : Two important databases, a directory database of current serials in Madras and a contents database covering articles published in 300 journals available in Madras libraries.

REFERENCES

1. http://www.webpages.uidaho.edu/~mbolin/jebaraj-networks.htm
2. www.inflibnet.ac.in
3. library.britishcouncil.org.in

4. www.mylibnet.org
5. delnet.nic.in
6. digitalcommons.unl.edu/cgi/viewcontent.cgi
7. www.eric.ed.gov/ERICWebPortal/recordDetail·
8. www.ingentaconnect.com/content/haworth/jildder/2006/00000016
9. www.icast.org.in/old/libnetwork.htm
10. www.ifla.org/IV/ifla66/papers/039-120e.htm
11. www.icast.org.in/old/libnetwork.htm
12. www.information-professional.netfirms.com/directory/lanind.htm

12
Metadata

Metadata (Greek: meta-+ Latin: data "information"), literally "data about data", is information that describes another set of data. In library routine environment, a common example is a library catalog card, which contains data about the contents and location of a book: it is data about the data in the book referred to by the card. Another important type of data about data is the links or relationship among data. Some metadata schemes attempt to embrace this concept such as Dublin Core element link. Since metadata is also data, it is possible to have "metadata of the metadata of data". However other machine generated data about data, such as the reversed index created by a free-text search engine is generally not considered as metadata. The metadata which is embedded with content is called *embedded metadata*. A data repository typically stores the metadata *detached* from the data. It can be further said that Metadata is machine understandable information for the web.

Technically, Metadata is structured data which describes the characteristics of a resource. It shares many similar characteristics to the cataloguing that takes place in libraries, museums and archives. The term "meta" derives from the Greek word denoting a nature of a higher order or more fundamental kind. Infact, a metadata record consists of a number of pre-defined elements representing specific attributes of a resource, and each element can have one or more values.

Each metadata schema usually has the following characteristics:

* a limited number of elements
* the name of each element
* the meaning of each element

Metadata includes control information and certain facts concerning an entry:

* automatic data (entry number, creator, Editor)
* general information (name, title, ordering, sequence no.),
* keywords,
* abstract and text,
* additional data.

TYPES OF METADATA

According to Library of Congress, Metadata elements are of three types: descriptive, administrative, and structural. Administrative metadata is used for managing and preserving objects in the repository; structural metadata is used primarily for storage of objects in a repository and for presentation; descriptive metadata is used for discovery of objects. The elements defined in this table are to support structural and administrative functions. Functions served are access management, administration, discovery, persistent identifier, presentation, digital preservation and preservation reformatting. Consequently, five descriptive elements are included in the Table of Core Metadata Elements.

LEVELS OF METADATA

According to Library of congress, levels of metadata are:

Set : Set-level metadata applies to what is currently known as a digital collection. Digital collections are formed from aggregates that group digital items by original content type, such as non-motion visual, text, motion visual, or audio; and by custodial responsibility as well as by collection. A collection may be determined from an archival series or from a topical bibliography. Some digital collections are contained within a single aggregate. Others are formed by numerous aggregates. Set-level metadata applies to all aggregates within the set regardless of content or responsibility.

Aggregate : An aggregate organizes digital objects by digital type and by digital custodial responsibility. A single aggregate may be a digital collection. Aggregate-level metadata applies to all primary objects within an aggregate.

Primary Object : The specific item identified by the online collection access aid as a coherent whole is known as the primary object. Primary objects are usually the digital equivalents of physical library items, such as a book, a sound recording, a movie, a single title of sheet music, a folder of letters, a photograph, or a map. This level metadata applies to all the intermediate and terminal objects of a particular primary object.

Intermediate Object : The intermediate object is a view or component of the primary object. A book that can be presented as page images or as searchable text has two intermediate objects. One points to all the page images of the book, the other to an encoded text file. Complex primary objects such as sound recordings offer many intermediate object possibilities. Multiple sides or tracks for a recording, each with sound and label components, may be captured in several formats. Metadata for an intermediate object allows the gathering of digital files and metadata for the creation of presentations.

Terminal Object : The terminal object is the digital content file or files that form the object. There is at least one terminal for each object. Terminal object-level metadata is primarily structural supplying the digital attributes of each file such as size, extension, bit-depth, etc

METADATA REGISTRY

A metadata registry is a central place where metadata definitions are stored and maintained.

A metadata registry typically has the following characteristics:

1. It is a protected area where only approved individuals may make changes
2. It stores Data elements that include both Semantics and Representations
3. The semantic areas of a metadata registry contain the meaning of a Data Element with precise definitions
4. The representational areas define how the data is represented in a specific format such as within a database or a structure file format such as XML

The International Organization for Standardization (ISO) has published standards for a metadata registry called ISO/IEC 11179. A metadata registry is frequently set up and administered by an organization's Data architect or data modeling team

Popular Metadata Schems

Some of the most popular metadata schemas include:

* Dublin Core
* AACR2 (Anglo-American Cataloging Rules)
* GILS (Government Information Locator Service)
* EAD (Encoded Archives Description)
* IMS (IMS Global Learning Consortium)
* AGLS (Australian Government Locator Service)

While the syntax is not strictly part of the metadata schema, the data will be unusable, unless the encoding scheme understands the semantics of the metadata schema. The encoding allows the metadata to be processed by a computer program. Important schemes include:

* HTML (Hyper-Text Markup Language)
* SGML (Standard Generalised Markup Language)
* XML (eXtensible Markup Language)
* RDF (Resource Description Framework)
* MARC (MAchine Readable Cataloging)

* MIME (Multipurpose Internet Mail Extensions)

Deployment of Metadata

Metadata may be deployed in a number of ways:

* Embedding the metadata in the Web page by the creator or their agent using META tags in the HTML coding of the page.
* As a separate HTML document linked to the resource it describes.
* In a database linked to the resource. The records may either have been directly created within the database or extracted from another source, such as Web pages.

The simplest method is for Web page creators to add the metadata as part of creating the page. Creating metadata directly in a database and linking it to the resource, is growing in popularity as an independent activity to the creation of the resources themselves. Increasingly, it is being created by an agent or third party, particularly to develop subject-based gateways.

Use of Metadata

Information resources must be made visible in a way that allows people to tell whether the resources are likely to be useful to them. This is very much important in the online world, and in particular, the World Wide Web. Metadata is a systematic method for describing resources and thereby improving access to them. Metadata maximize the ability to locate it. Metadata provides the essential link between the information creator and the information user. While the primary aim of metadata is to improve resource discovery, metadata sets are also being developed for several other reasons, including:

* management information
* content rating
* rights management
* administrative control
* security

* personal information
* preservation

ELEMENTS IN DUBLIN CORE

Elements in Dublin Core are given as below. The metadata elements fall into three groups which roughly indicate the class or scope of information stored in them: (1) elements related mainly to the content of the resource, (2) elements related mainly to the resource when viewed as intellectual property, and (3) elements related mainly to the physical manifestation of the resource.

Content & about the Resource	*Intellectual Property*	*Electronic or Physical manifestation*
Title	Author or Creator	Date
Subject	Publisher	Type
Description	Contributor	Format
Source	Rights	Identifier
Language		
Relation		
Coverage		

Here is an example of a Dublin Core record for a short poem, encoded as part of a Web page using the <META> tag:

<HTML> !4.0!

<HEAD>

<TITLE>Song</TITLE>

<META NAME="DC.Title" CONTENT="Song"

<META NAME="DC.Creator" CONTENT="Nash, Ogden">

<META NAME="DC.Type" CONTENT="text">

<META NAME="DC.Date" CONTENT="1939">

<META NAME="DC.Format" CONTENT="text/html">

<META NAME="DC.Identifier" CONTENT="http://www.epoetry.com/beauty.html">

```
</HEAD>
<BODY><PRE>
she walks in beauty
</PRE></BODY>
</HTML>
```

The <META> tag is not normally displayed by Web browsers, but can be viewed by selecting "Page Source".

In addition to the 15 elements, three qualifying aspects have been accepted to enable the Dublin Core to function in an international context and also meet higher level scientific and subject-specific resource discovery needs. These three Dublin Core Qualifiers are:

* **LANG :** indicating the language of the contents of the element, to be used in both resource discovery and in filtering retrieval results
* **SCHEME :** indicating the set of regulations, standards, conventions or norms from which a term in the content of the element has been taken
* **SUB-ELEMENT :** refinement of some of the elements to gain more precision

USE OF ELEMENTS, SUB-ELEMENTS AND SCHEMES

Use of elements, sub-elements and schemes is based on:

* the specific needs of the local community to maximize information retrieval and management
* the need to guard against making the creation of metadata and its maintenance more trouble than it is worth and therefore defeating its purpose
* sustainability of the metadata schema in terms of keeping the records up to date

The important aspect is that a simple description is better than no description at all, as long as it can aid in the consistent discovery of resources. The level of specificity in resource description is also important. The resources can be described individually or at a collection or aggregate level. It would be practically impossible to provide guidelines as to the appropriate level of specificity. Initially, it would be sensible to allow the creators to determine which resources deserve their own record. If a collection-level record is used, it is important to add as much information as possible to ensure appropriate retrieval. Acting on users' feedback is also important. Monitoring the search terms input by customers, is an important technique for improving the quality and coverage of a database.

CREATING METADATA

The metadata can be created and collected easily at point of creation of a resource or at point of publication. There are many such tools which are either standalone or part of a package of software, usually with a backend database or repository to store and retrieve the metadata records, Some examples include:

* DC-dot - http://www.ukoln.ac.uk/metadata/dcdot/. This service will retrieve a Web page and automatically generate Dublin Core metadata, either as HTML tags or as RDF/XML, suitable for embedding in the section of the page.
* DCmeta - http://www.dstc.edu.au/RDU/MetaWeb/generic_tool.html. Developed by Tasmania Online. It is based on SuperNoteTab text-editor and can be customised.
* HotMeta -. A package of software, including metadata editor, repository and search engine.

Ideally, metadata is created using a purpose-built tool, with the manual creation of data kept to an absolute minimum. The tool should support:

* Inclusion of the syntax in the template (e.g. element name, sub-element, qualifier)
* Default content, which can be overridden
* Content selected from a list of limited choices (e.g. Function, Type, Format)
* Validation of mandatory elements, sub-elements, schemes

and element values

Metadata and Resource Description at World Wide Web Consortium (W3C)

Metadata is machine understandable information for the web. The W3C Metadata Activity addressed the combined needs of several groups for a common framework to express assertions about information on the Web, and was superceded by the W3C Semantic Web Activity. W3C's Metadata Activity addressed the needs of PICS, DSig, P3P, and CC/PP. The primary work in this activity was the Resource Description Framework.

W3C Metadata Activity

There is now a wealth of information on every subject available on the Net. Today's Web gives people access to news, to the weather and to financial services. Important is a part of the Web which contains *information about information* i.e., labeling, cataloging and descriptive information structured in such a way that allows Web pages to be properly searched and processed in particular *by computer*. In other words, metadata is needed on the Web. W3C's Metadata Activity is concerned with ways to model and encode metadata, with a particular priority to use the Web to document the *meaning* of the metadata.

THE RESOURCE DESCRIPTION FRAMEWORK - RDF

The development of the Resource Description Framework (RDF) took, which provides a more general treatment of metadata. RDF is a declarative language and provides a standard way for using XML to represent metadata in the form of statements about properties and relationships of items on the Web. Such items, known as *resources*, can be almost anything, provided it has a Web address. This means that one can associate metadata with a Web page, a graphic, an audio file, a movie clip, and so on.

Practical uses of RDF

There are many practical uses of RDF, described below :

Thesauri and library classification schemes : These are well known examples of hierarchical systems for representing subject taxonomies in terms of the relationships between named concepts. The RDF Schema specification has exactly the features for creating RDF models that represent the logical structure of thesauri and other library classification systems.

Web sitemaps : A sitemap can be seen "internally" as a description of a Web site. The RDF Schema specification provides a mechanism for defining the vocabulary needed for this kind of application.

Description of the contents of Web pages : This is one of the basic functions of the Dublin Core initiative. The Dublin Core is a set of 15 properties associated with bibliographic information. These can be used to describe items on the Web sufficiently well that search engines and other software can work much more efficiently.

Describing the formal structure of privacy practice descriptions : W3C's Platform for Privacy Preferences Project (P3P) is working on a platform that allows users to be informed of a site's practices. Users or software operating on their behalf, can then negotiate for a different privacy policy and come to an agreement with the site which will be the basis for any subsequent release of information. RDF may be used to describe the formal structure of privacy practice descriptions.

Descriptions of device capabilities : The Mobile Access Activity has an interest in a framework for describing the display and processing capabilities of mobile Web devices. RDF provides a way to describe the capabilities and preferences associated with users and the hardware and software they are using to access the Web.

Rating systems : These offer a way of labeling resources so that people (or computers) can filter information. RDF enables programmers to devise rating systems for any number of domains.

Expressing metadata about metadata : One can also use

RDF to describe metadata about a given rating - the date it was given, by which organization, and so on.

Digital Signatures : As the Web matures and more everyday tasks are performed online, so *digital signatures* are become increasingly important. RDF may be used to express information concerning what users are signing, what the significance of the signature is, and the dates that the signature is valid, and so on.

W3C Semantic Web Activity

The **Semantic Web** provides a common framework that allows **data** to be shared and reused across application, enterprise, and community boundaries. It is a collaborative effort led by W3C with participation from a large number of researchers and industrial partners. It is also based on the Resource Description Framework (RDF). The Semantic Web is a web of data. There is lots of data we all use every day, and it's not part of the web. The Semantic Web is about two things. It is about common formats for integration and combination of data drawn from diverse sources, where on the original Web mainly concentrated on the interchange of documents. It is also about language for recording how the data relates to real world objects. That allows a person, or a machine, to start off in one database, and then move through an unending set of databases which are connected not by wires but by being about the same thing.

THE DUBLIN CORE METADATA INITIATIVE

The Dublin Core Metadata Initiative (DCMI) is an open organization engaged in the development of interoperable online metadata standards that support a broad range of purposes and business models. DCMI's activities include work on architecture and modeling, discussions and collaborative work in DCMI Communities and DCMI Task Groups, annual conferences and workshops, standards liaison, and educational efforts to promote widespread acceptance of metadata standards and practices

DCMI Metadata Terms

Each term is specified with the following minimal set of attributes:

Name : The unique token assigned to the term.

URI : The Uniform Resource Identifier used to uniquely identify a term.

Label : The human-readable label assigned to the term.

Definition : A statement that represents the concept and essential nature of the term.

Type of Term : The type of term, such as Element or Encoding Scheme, as described in the DCMI Grammatical Principles.

Status : Status assigned to term by the DCMI Usage Board, as described in the DCMI Usage Board Process.

Date issued : Date on which a term was first declared.

Where applicable, the following attributes provide additional information about a term:

Comment : Additional information about the term or its application.

See : A link to authoritative documentation.

References : A citation or URL of a resource referenced in the Definition or Comment.

Refines : A reference to a term refined by an Element Refinement.

Qualifies : A reference to a term qualified by an Encoding Scheme.

Broader Than : A reference from a more general to a more specific Vocabulary Term.

Narrower Than : A reference from a more specific to a more general Vocabulary Term.

THE DUBLIN CORE METADATA STANDARD

The Dublin Core metadata standard is a simple yet effective element set for describing a wide range of networked resources. The Dublin Core standard includes two levels: Simple and Qualified. Simple Dublin Core comprises fifteen elements; Qualified Dublin Core includes three additional elements (Audience, Provenance and RightsHolder), as well as a group of element refinements (also called qualifiers) that refine the semantics of the elements in ways that may be useful in resource discovery. The semantics of Dublin Core have been established by an international, cross-disciplinary group of professionals from librarianship, computer science, text encoding, the museum community, and other related fields of scholarship and practice.

Another way to look at Dublin Core is as a "small language for making a particular class of statements about resources". In this language, there are two classes of terms — elements (nouns) and qualifiers (adjectives) — which can be arranged into a simple pattern of statements. The resources themselves are the implied subjects in this language. In the diverse world of the Internet, Dublin Core can be seen as a "metadata pidgin for digital tourists": easily grasped, but not necessarily up to the task of expressing complex relationships or concepts.

The Dublin Core basic element set is outlined. Each element is optional and may be repeated. Most elements also have a limited set of qualifiers or refinements, attributes that may be used to further refine (not extend) the meaning of the element. The Dublin Core Metadata Initiative (DCMI) has established standard ways to refine elements and encourage the use of encoding and vocabulary schemes. The full set of elements and element refinements conforming to DCMI "best practice" is available, with a formal registry available as well.

Three other Dublin Core principles are as following as they are critical to understanding how to think about the relationship of metadata to the underlying resources:

1. The One-to-One Principle : In general Dublin Core metadata describes one manifestation or version of a resource, rather than assuming that manifestations stand in for one another. The relationship between the metadata for the original and the reproduction is part of the metadata description, and assists the user in determining whether he or she needs to go to the Louvre for the original, or whether his/her need can be met by a reproduction.

2. The Dumb-down Principle : The qualification of Dublin Core properties is guided by a rule known colloquially as the Dumb-Down Principle. According to this rule, a client should be able to ignore any qualifier and use the value as if it were unqualified. While this may result in some loss of specificity, the remaining element value (minus the qualifier) must continue to be generally correct and useful for discovery. Qualification is therefore supposed only to refine, not extend the semantic scope of a property.

3. Appropriate values : Best practice for a particular element or qualifier may vary by context, but in general an implementor cannot predict that the interpreter of the metadata will always be a machine. This may impose certain constraints on how metadata is constructed, but the requirement of usefulness for discovery should be kept in mind.

Although the Dublin Core was originally developed with an eye to describing document-like objects (because traditional text resources are fairly well understood), DC metadata can be applied to other resources as well. Its suitability for use with particular non-document resources will depend to some extent on how closely their metadata resembles typical document metadata and also what purpose the metadata is intended to serve.

Goals of Dublin Core : Dublin Core has as its goals:

Simplicity of Creation and Maintenance

The Dublin Core element set has been kept as small and simple as possible to allow a non-specialist to create simple descriptive records for information resources easily and inexpensively, while providing for effective retrieval of those resources in the networked environment.

Commonly Understood Semantics

Discovery of information across the vast commons of the Internet is hindered by differences in terminology and descriptive practices from one field of knowledge to the next. The Dublin Core can help the "digital tourist" — a non-specialist searcher — find his or her way by supporting a common set of elements, the semantics of which are universally understood and supported.

International Scope

The Dublin Core Element Set was originally developed in English, but versions are being created in many other languages, including Finnish, Norwegian, Thai, Japanese, French, Portuguese, German, Greek, Indonesian, and Spanish. The DCMI Localization and Internationalization Special Interest Group is coordinating efforts to link these versions in a distributed registry.

Although the technical challenges of internationalization on the World Wide Web have not been directly addressed by the Dublin Core development community, the involvement of representatives from virtually every continent has ensured that the development of the standard considers the multilingual and multicultural nature of the electronic information universe. While balancing the needs for simplicity in describing digital resources with the need for precise retrieval, Dublin Core developers have recognized the importance of providing a mechanism for extending the DC element set for additional resource discovery needs. It is expected that other communities of metadata experts will create and administer additional metadata sets, specialized to the needs of their communities. Metadata elements from these sets could be used in conjunction with Dublin Core metadata to meet the need for interoperabilbility. The DCMI Usage Board is presently working on a model for accomplishing this in the context of "application profiles."

Rachel Heery and Manjula Patel, in their article

"Application profiles: mixing and matching metadata schemas" define an application profile as:

" ... schemas which consist of data elements drawn from one or more namespaces, combined together by implementors, and optimised for a particular local application."

This model allows different communities to use the DC elements for core descriptive information, and allowing domain specific extensions which make sense within a more limited arena.

DUBLIN CORE METADATA SCHEMA

Element	*Element description*
Creator	Person or organisation primarily responsible for creating the intellectual content of the resource, e.g. authors in the case of written documents, artists, photographers, etc. in the case of visual resources.
Publisher	The entity (e.g. agency including unit/branch/section) responsible for making the resource available in its present form, such as a publishing house, a university department, or a corporate entity.
Contributor	Person or organisation not specified in a Creator element who has made significant intellectual contributions to the resource but whose contribution is secondary to any person or organisation specified in a Creator element, e.g. editor, transcriber, illustrator.
Rights Management	A rights management statement, an identifier that links to a rights management statement.

Title	The name given to the resource, usually by the creator or publisher. Can be the same as the title of the resource, or may be more descriptive
Subject	The topic of the resource. Typically, will be expressed as keywords or phrases that describe the subject or content of the resource. Controlled vocabularies and formal classification schemes are encouraged.
Date	A date associated with the creation or availability of the resource.
Identifier	A string or number used to uniquely identify the resource. Examples for networked resources include URLs, Purls and URNs. ISBN or other formal names can be used.
Description	A textual description of the content of the resource, including abstracts in the case of document-like objects or content descriptions in the case of visual resources.
Source	The work, either print or electronic, from which this object is derived, if applicable. Source is not applicable if the present resource is in its original form.
Language	The language of the intellectual content of the resource.
Relation	Relationship to other resources, e.g. images in a document, chapters in a book, items in a collection.
Coverage	Spatial locations and temporal duration characteristic of the resource.

Type	The category of the resource, such as home page, novel, poem, working paper, technical report, essay, dictionary.
Format	The data format of the resource, used to identify the software and possibly hardware that might be needed to display or operate the resource, e.g. postscript, HTML, text, jpeg, XML.

ELEMENT CONTENT AND CONTROLLED VOCABULARIES

Each Dublin Core element is optional and repeatable, and there is no defined order of elements. The ordering of multiple occurrences of the same element (e.g., Creator) may have a significance intended by the provider, but ordering is not guaranteed to be preserved in every user environment. Ordering or sequencing may be syntax dependent; for instance, RDF/ XML supports ordering, but HTML does not.

Content data for some elements may be selected from a "controlled vocabulary," which is a limited set of consistently used and carefully defined terms. This can improve search results because computers are good at matching words character by character but weak at understanding the way people refer to one concept using different words, i.e. synonyms. Without basic terminology control, inconsistent or incorrect metadata can profoundly degrade the quality of search results. Controlled vocabularies may also reduce the likelihood of spelling errors when recording metadata.

Two costs are associated with controlled vocabulary. One cost of a controlled vocabulary is the necessity for an administrative body to review, update and disseminate the vocabulary. For example, the US Library of Congress Subject Headings (LCSH) and the US National Library of Medicine Medical Subject Headings (MeSH) are formal vocabularies, indispensable for searching rigorously cataloged collections. However, both require significant support organizations. Another cost is having to train searchers and creators of metadata so that they know when using MeSH, for example, to enter "myocardial infarction" instead of the more colloquial "heart attack." More sophisticated implementations can make such tasks much easier, but the controlled vocabulary terms must be available for them to apply.

USING DUBLIN CORE - THE ELEMENTS

1. Title

Label: Title

Element Description : The name given to the resource. Typically, a Title will be a name by which the resource is formally known.

While not in the position of determining the title, the Title element can be repeated and the variants can be included in second and subsequent Title iterations. If the item is in HTML, the source document is to be viewed and it can be made sure that the title identified in the title header (if any) is also included as a Title.

2. Subject

Label: Subject and Keywords

Element Description: The topic of the content of the resource. Typically, a Subject is expressed as keywords or key phrases or classification codes that describe the topic of the resource.

Subject keywords from the Title or Description information, or from within a text resource are selected. If the subject of the item is a person or an organization, the same form of the name will be used as if the person or organization were a Creator or Contributor. In general, the most significant and unique words are chosen for keywords, avoiding those too general to describe a particular item. Subject might include classification data if it is available or controlled vocabularies as well as keywords. When including terms from multiple vocabularies, separate element iterations are used.

3. Description

Label: Description

Element Description : An account of the content of the resource. Description may include but is not limited to: an abstract, table of contents, reference to a graphical representation of content or a free-text account of the content.

Since the Description field is a potentially rich source of indexable terms, care should be taken to provide this element when possible. Descriptive information can be copied or automatically extracted from the item if there is no abstract or other structured description available. Although the source of the description may be a web page or other structured text with presentation tags, it is generally not good practice to include HTML or other structural tags within the Description element. Applications vary considerably in their ability to interpret such tags, and their inclusion may negatively affect the interoperability of the metadata.

4. Type

Label : Resource Type

Element Description : The nature or genre of the content of the resource. Type includes terms describing general categories, functions, genres, or aggregation levels for content. Recommended best practice is to select a value from a controlled vocabulary. To describe the physical or digital manifestation of the resource, FORMAT element is used.

If the resource is composed of multiple mixed types then multiple or repeated Type elements should be used to describe the main components. Because different communities or domains are expected to use a variety of type vocabularies, best practice to ensure interoperability is to include at least one general type term from the DCMIType vocabulary in addition to the domain specific type term(s), in separate Type element iterations.

5. Source

Label : Source

Element Description : A Reference to a resource from which the present resource is derived. The present resource may be derived from the Source resource in whole or part.

In general, in this area information about a resource is included that is related intellectually to the described resource but does not fit easily into a Relation element.

6. Relation

Label: Relation

Element Description : A reference to a related resource. Recommended best practice is to reference the resource by means of a string or number conforming to a formal identification system.

Relationships may be expressed reciprocally (if the resources on both ends of the relationship are being described) or in one direction only, even when there is a refinement available to allow reciprocity. If text strings are used instead of identifying numbers, the reference should be appropriately specific. For instance, a formal bibliographic citation might be used to point users to a particular resource. Because the refined terms used with Relation provide significantly more information to a user than the unqualified use of Relation, implementers who are describing heavily interrelated resources might choose to use qualified Dublin Core.

7. Coverage

Label: Coverage

Element Description : The extent or scope of the content of the resource. Coverage will typically include spatial location (a place name or geographic co-ordinates), temporal period (a period label, date, or date range) or jurisdiction (such as a named administrative entity). Where appropriate, named places or time periods should be used in preference to numeric identifiers such as sets of co-ordinates or date ranges.

Whether this element is used for spatial or temporal information, care should be taken to provide consistent information that can be interpreted by human users, particularly in order to provide interoperability in situations where sophisticated geographic or time-specific searching is not supported. For complex applications, consideration should be given to using an encoding scheme that supports appropriate specification of information, such as DCMI Period, DCMI Box or DCMI Point.

8. Creator

Label: Creator

Element Description : An entity primarily responsible for making the content of the resource. Examples of a Creator include a person, an organization, or a service. Typically the name of the Creator should be used to indicate the entity.

Creators should be listed separately, preferably in the same order that they appear in the publication. Personal names should be listed surname or family name first, followed by forename or given name. When in doubt, give the name as it appears, and do not invert. In the case of organizations where there is clearly a hierarchy present, the parts of the hierarchy from largest to smallest are listed, separated by full stops and a space. If it is not clear whether there is a hierarchy present, or unclear which is the larger or smaller portion of the body, the name are given as it appears in the item. If the Creator and Publisher are the same, do not repeat the name in the Publisher area. If the nature of the responsibility is ambiguous, the recommended practice is to use Publisher for organizations, and Creator for individuals. In cases of lesser or ambiguous responsibility, other than creation, use Contributor.

9. Publisher

Label : Publisher

Element Description : The entity responsible for making the resource available. Examples of a Publisher include a person, an organization, or a service. Typically, the name of a Publisher should be used to indicate the entity.

The intent of specifying this field is to identify the entity that provides access to the resource. If the Creator and Publisher are the same, do not repeat the name in the Publisher area. If the nature of the responsibility is ambiguous, the recommended practice is to use Publisher for organizations, and Creator for individuals. In cases of ambiguous responsibility, use Contributor.

10. Contributor

Label : Contributor

Element Description : An entity responsible for making contributions to the content of the resource. Examples of a Contributor include a person, an organization or a service. Typically, the name of a Contributor should be used to indicate the entity.

The general guidelines for using names of persons or organizations as Creators apply here. Contributor is the most general of the elements used for "agents" responsible for the resource, so should be used when primary responsibility is unknown or irrelevant.

11. Rights

Label : Rights Management

Element Description : Information about rights held in and over the resource. Typically a Rights element will contain a rights management statement for the resource, or reference a service providing such information. Rights information often encompasses Intellectual Property Rights (IPR), Copyright, and various Property Rights. If the rights element is absent, no assumptions can be made about the status of these and other rights with respect to the resource.

12. Date

Label : Date

Element Description : A date associated with an event in the life cycle of the resource. Typically, Date will be associated with the creation or availability of the resource. If the full date is unknown, month and year (YYYY-MM) or just year (YYYY) may be used. Many other schemes are possible, but if used, they may not be easily interpreted by users or software.

13. Format

Label : Format

Element Description : The physical or digital manifestation of the resource. Typically, Format may include the media-type or dimensions of the resource. Examples of dimensions include size and duration. Format may be used to determine the software, hardware or other equipment needed to display or operate the resource.

In addition to the specific physical or electronic media format, information concerning the size of a resource may be included in the content of the Format element if available. In resource discovery size, extent or medium of the resource might be used as a criterion to select resources of interest, since a user may need to evaluate whether they can make use of the resource within the infrastructure available to them. When more than one category of format information is included in a single record, they should go in separate iterations of the element.

14. Identifier

Label : Resource Identifier

Element Description : It is an unambiguous reference to the resource within a given context. Examples of formal identification systems include the Uniform Resource Identifier (URI) (including the Uniform Resource Locator (URL), the Digital Object Identifier (DOI) and the International Standard Book Number (ISBN).

This element can also be used for local identifiers (e.g.

ID numbers or call numbers) assigned by the Creator of the resource to apply to a particular item. It should not be used for identification of the metadata record itself.

15. Language

Label : Language

Element Description : A language of the intellectual content of the resource. The best practice for the values of the Language element is defined by RFC 3066 [RFC 3066, http://www.ietf.org/rfc/ rfc3066.txt] which, in conjunction with ISO 639 [ISO 639, http://www.oasis- open.org/cover/iso639a.html]), defines two- and three-letter primary language tags with optional subtags. Examples include "en" or "eng" for English, "akk" for Akkadian, and "en-GB" for English used in the United Kingdom. Either a coded value or text string is represented here.

16. Audience

Label : Audience

Element Description : A class of entity for whom the resource is intended or useful. A class of entity may be determined by the creator or the publisher or by a third party.

Audience terms are best utilized in the context of formal or informal controlled vocabularies. None arc presently recommended or registered by DCMI, but several communities of interest are engaged in setting up audience vocabularies. In the absence of recommended controlled vocabularies, implementors are encouraged to develop local lists of values, and to use them consistently.

17. Provenance

Label : Provenance

Element Description : A statement of any changes in ownership and custody of the resource since its creation that are significant for its authenticity, integrity and interpretation. The statement may include a description of any changes successive custodians made to the resource.

18. RightsHolder

Label : Rights Holder

Element Description : A person or organization owning or managing rights over the resource. It is to use the URI (Uniform Resource Identifier) or name of the Rights Holder to indicate the entity.

Since, for the most part, people and organizations are not typically assigned URIs, a person or organization holding rights over a resource would be named using a text string. People and organizations sometimes have websites, but URLs for these are not generally appropriate for use in this context, since they are not clearly identifying the person or organization, but rather the location of a website about them.

19. InstructionalMethod

Label : Instructional Method

Element Description : A process, used to engender knowledge, attitudes and skills, that the resource is designed to support. Instructional Method will typically include ways of presenting instructional materials or conducting instructional activities, patterns of learner-to-learner and learner-to-instructor interactions, and mechanisms by which group and individual levels of learning are measured. Instructional methods include all aspects of the instruction and learning processes from planning and implementation through evaluation and feedback.

20. AccrualMethod

Label : Accrual Method

Element Description : The method by which items are added to a collection.

Terms from controlled vocabularies may be developed for the use of a particular project or in general use in a cultural materials context.

21. AccrualPeriodicity

Label : Accrual Periodicity

Element Description : The frequency with which items are added to a collection.

Terms from controlled vocabularies may be developed for the use of a particular project or in general use in a cultural materials context.

22. AccrualPolicy

Label : Accrual Policy

Element Description : The policy governing the addition of items to a collection.

DUBLIN CORE QUALIFIERS

The DCMI recognizes two broad classes of qualifiers:

Element Refinement : These qualifiers make the meaning of an element narrower or more specific. A refined element shares the meaning of the unqualified element, but with a more restricted scope. A client that does not understand a specific element refinement term should be able to ignore the qualifier and treat the metadata value as if it were an unqualified (broader) element. The definitions of element refinement terms for qualifiers must be publicly available.

Encoding Scheme : These qualifiers identify schemes that aid in the interpretation of an element value. These schemes include controlled vocabularies and formal notations or parsing rules. A value expressed using an encoding scheme will thus be a token selected from a controlled vocabulary (e.g., a term from a classification system or set of subject headings) or a string formatted in accordance with a formal notation. If an encoding scheme is not understood by a client or agent, the value may still be useful to a human reader. The definitive description of an encoding scheme for qualifiers must be clearly identified and available for public use.

PROPERTIES OF DUBLIN CORE QUALIFIERS

Dublin Core qualifiers have the following properties:

* **Name :** The unique token assigned to the qualifier.

* **Label :** The human-readable label assigned to the qualifier.
* **Definition :** A statement that represents the concept and essential nature of the qualifier.
* **Comment :** Additional information associated with the qualifier (if available).
* **See Also :** A link to more information about the qualifier (if available).

Importance of Dublin Core

The Dublin Core metadata schema offers the following advantages:

* Its usability and its flexibility
* The semantics of these elements is designed to be clear enough to be understood by a wide range of customers, without the need for training
* The elements of Dublin Core are easily identifiable by having the work in hand, such as intellectual content and physical format
* It is intended to supplant other resource descriptions as well as to complement them. It is intended to describe the essential features of electronic documents that support resource discovery.
* It is mostly syntax independent, to support its use in the widest range of applications
* All elements are optional, but allows each site to define which elements are mandatory and which are optional
* All elements are repeatable
* The elements may be modified in limited and well-defined ways through the use of specific qualifiers, such as the name of the thesaurus used in the subject element
* From the very beginning, the Dublin Core creators recognized that some resources could not be adequately described by a small set of elements. The Dublin Core creators came up with two solutions. Firstly, by allowing

the addition of elements for site-specific purposes or specialized fields. Secondly, by designing the Dublin Core schema so that it could be mapped into more complex and tightly controlled systems, such as MARC.

Dublin Core has received widespread acceptance amongst the resource discovery community and has become the de facto Internet metadata standard. To date, the depth of implementation in individual sectors has been patchy.

Important instructions about Metadata Creation

Keeping metatags simple and concise : Very long descriptions and too numerous keywords will be ignored by some spiders, e.g. some search engines index only the first six keywords. Search engines can also detect metatag spamming, where a keyword is repeated several times, and may penalize a site's ranking.

Prioritizing keywords : Since some search engines only read a certain number of keywords it is important to list the most important ones first.

Avoiding the REFRESH metatag for re-direction : Although a very handy tool, so-called bridge pages should be avoided because they tend to clutter search engine's databases and confuse users.

IMPORTANCE FOR LIBRARY

According to library o Congress, METS (Metadata Encoding & Transmission Standards) maintaining a library of digital objects of necessity requires maintaining metadata about those objects. The metadata is necessary for successful management and use of digital objects is both more extensive than and different from the metadata used for managing collections of printed works and other physical materials. While a library may record descriptive metadata regarding a book in its collection, the book will not dissolve into a series of unconnected pages if the library fails to record structural metadata regarding the book's organization, nor will scholars be unable to evaluate the book's worth if the library fails to note that the book was produced using a particular offset press. The same cannot be said for a digital version of the same book. Without structural metadata, the page image or text files comprising the digital work are of little use, and without technical metadata regarding the digitization process, scholars may be unsure of how accurate a reflection of the original the digital version provides. For internal management purposes, a library must have access to appropriate technical metadata in order to periodically refresh and migrate the data, ensuring the durability of valuable resources. Metatags are, as netmechanic calls them, "the librarian's favorite tag". Metatags present an opportunity to index and catalogue this vast new resource more effectively. Metatag syntax has already been adapted to several subject field and languages and many projects are working on developing standards and software tools.

REFERENCES

1. [AGLS] Australian Government Locator Service Implementation Plan: A Report by the Australian Government Locator Service Working Party (AGLS WG) December, 1997.
2. [AGLS1] AGLS Metadata Element Set. National Archives of Australia. http://www.naa.gov.au/recordkeeping/gov_online/agls/metadata_element_set.html
3. [DC1] The Essential Elements of Networked Object Description. Stuart Weibel. OCLC/NCSA Metadata Workshop, March, 1995. http://www.oclc.org:5046/oclc/research/metadata/dublin_core_report.html
4. [DESIRE] Specification for resource description methods Part 1: A review of metadata: a survey of current resource description formats. Lorcan Dempsey and Rachel Heery, March,1997. http://www.ukoln.ac.uk/metadata/desire/overview/
5. [EC] Metadata Workshop. European Commission, Telematics for Libraries, December, 1997. http://hosted.ukoln.ac.uk/ec/metadata-1997/
6. [LC] The Library of Congress Portals Applications Issues Group.http://www.loc.gov/catdir/lcpaig/
7. [NISO] Information Retrieval (Z39.50) - Application Service: Definition and Protocol Specification (Version 3). The National

Information Standards Organization, 1995. http://www.niso.org/

8. [OAI] Open Archives Initiative. http://www.openarchives.org/
9. [SEW] Search Engine Watch. Search Engine Feature Company. http://searchenginewatch.com/webmasters/features.html
10. [WGGIN] Improving Access to Information and Services of Australian Governments. Working Group on Government Information Navigation, July, 1997. http://www.nla.gov.au/lis/esd4.html
11. http://www.dstc.edu.au/Research/Projects/hotmeta/
12. http://dublincore.org/documents/dcmi-terms/
13. introduction to Metadata http://www.library.uq.edu.au/iad/ctmeta4.html
14. Metadata and Resource Description http://www.w3.org/Metadata/
15. http://www.loc.gov/standards/mets/METSOVERVIEW.V2.html
16. http://en.wikipedia.org/wiki/metadata

13
Artificial Intelligence

Artificial intelligence as a separate discipline started in the 1950s when it was recognised that computers were not just giant calculators dealing with numbers, but logic machine that could process symbols, expressed as numbers, letters of the alphabet, or words in a language. Since then it has grown rapidly and today it has attained the status of a full discipline. There are unmistakable sign that interest in AI us increasing steadily; universally. Basically AI is the study of ways in which computers can be made to perform cognitive tasks, at which at present people are better. Examples of problems that fall under the aegis of AI include common-sense tasks, such as understanding English, recognizing scenes of the plot of a mystery novel.

Both the scientific and the engineering disciplines of AI are based on the physical symbol system hypothesis, which states that the processes that are required to produce intelligent action can be simulated with a collection of physical symbols and a set of mechanism that produce a series, over time of structures built from these symbols. The digital computer serves as tool with which the symbol structures are built and manipulated. In AI program symbol structures are used to represent both general knowledge about a problem domain and specific knowledge about the solution to the current problem.

DEFINING ARTIFICIAL INTELLIGENCE

Many definitions are floating about AI based on their functions etc. some definitions are as following:

1. AI is the study of ways to make computers be intelligent.
2. It is an attempt to make a computer respond like a human being.
3. AI is the ability of a computer to resolve uncertainty through what-ever processes it has available it.
4. AI is the science of modeling human intelligence.
5. AI is the part of the computer science that is systems that exhibit the characteristics we associate with intelligence in human behaviour- understanding language, learning, reasoning, solving problems and so on.

Really intelligent applications will only be achievable with artificial intelligence and it is the mark of a successful AI software designer to deliver functionality that can't be delivered without using AI. Computers like things precise: on or off, one or zero, yes or no. Artificial intelligence is about making computers act more like humans and one might be surprised at how many places it's showing up - from cameras and fridges to spam filters. Researchers are still arguing about whether a computer could actually think or just respond as if it can, but the results of their research are too useful to stay in the lab. Software can act in ways we think of as intelligent, can deliver results we'd only expect from a human, and can let us work in ways that feel more natural.

Artificial intelligence (also some times called Synthetic Intelligence) is defined as intelligence exhibited by an artificial entity. Such a system is generally assumed to be a computer. What the field of AI is really about is inventing machines that will help people in a variety of ways, by giving machines some of the sophisticated capabilities that humans have, such as the ability to understand spoken words, or interpret images, or to learn from experience.

MEANING OF ARTIFICIAL INTELLIGENCE

Artificial intelligence can mean many things to many people. Much confusion arises because the word 'intelligence' is ill-defined. The phrase is so broad that people have found it useful to divide AI. It has been divided as strong AI and weak AI. Strong AI claims that computers can be made to think on a level (at least) equal to humans and possibly even be conscious of themselves. Weak AI simply states that some "thinking-like" features can be added to computers to make them more useful tools and this has already started to happen. What does 'think' and 'thinking-like' mean? That's a matter of much debate.

In practice AI tends not to be simple algorithms that are straightforward preprogrammed sequences of commands. There are other gray areas too. For example, some people would say that Google is AI and some people would call it database. Some people would say its algorithm or theoretical computer science. Infact it contains some elements of all of the above. It is interesting to discern, where is the dividing line between all these. It is the science and engineering of making intelligent machines, especially intelligent computer programs. Intelligence is the computational part of the ability to achieve goals in the world.

Types of AI

AI research and developments are largely based on existing hardware and operating systems. The mechanisms for achieving intelligent behavior are not at all well understood, researchers need experimental environments that are easy to work in. It is assumed that once some of the mechanisms are worked out at a conceptual level it will be possible to optimize them by mapping them into hardware or systems capabilities.

Till now, two types of AI are categorized:

1. Conventional AI and
2. Computational AI (CI).

Infact, conventional AI mostly involves methods now classified as machine learning, characterized by formalism and statistical analysis. This is also known as symbolic AI, logical AI, neat AI and Good Old Fashioned Artificial Intelligence (GOFAI). The methods include:

* **Expert systems :** apply reasoning capabilities to reach a conclusion. An expert system can process large amounts of known information and provide conclusions based on them.
* Case based reasoning
* **Behavior based AI :** a modular method of building AI systems by hand.

Computational Intelligence involves iterative development or learning. Learning is based on empirical data and is associated with non-symbolic AI, scruffy AI and soft computing. Methods mainly include:

* **Neural networks :** these are systems with very strong pattern recognition capabilities.
* **Fuzzy systems :** these techniques for reasoning under uncertainty, has been widely used in modern industrial and consumer product control systems.
* **Evolutionary computation :** it applies biologically inspired concepts such as populations, mutation and survival of the fittest to generate increasingly better solutions to the problem. These methods most notably divide into evolutionary algorithms (e.g. genetic algorithms) and swarm intelligence (e.g. ant algorithms).

Statistical AI, arising from machine learning, tends to be more concerned with "inductive" thought: given a set of patterns, it induces the trend. Classical AI, on the other hand, is more concerned with "deductive" thought: given a set of constraints, it deduces a conclusion. Another difference is found that C++ tends to be a favorite language for statistical AI while LISP dominates in classical AI.

Branches of AI

AI is found in various forms. Important branches are explained as following :

logical AI

In the logical AI, the program decides what to do by inferring that certain actions are appropriate for achieving its objectives.

search

AI programs often examine large numbers of possibilities, e.g. moves in a chess game or inferences by a theorem proving program. Discoveries are continually made about how to do this more efficiently in various domains.

pattern recognition

While observing, the programs compare what it sees with a pattern. For example, a vision program may try to match a pattern of eyes and a nose in a scene in order to find a face. More complex patterns, e.g. in a natural language text or in a chess position are also studied. The complex patterns require quite different methods than do the simple patterns that have been studied the most.

representation

For representation, usually languages of mathematical logic are used in the AI.

inference

In the intelligence, from some facts, others can be inferred. Mathematical logical deduction is adequate for some purposes, but new methods of *non-monotonic* inference have been added to logic since the 1970s. For example, when we hear of a bird, we infer that it can fly, but this conclusion can be reversed if it is a penguin. Ordinary logical reasoning is monotonic in that the set of conclusions that can the drawn from a set of premises is a monotonic increasing function of the premises.

common sense knowledge and reasoning

In this area AI is farthest from human-level. There has been considerable progress, e.g. in developing systems of *non-monotonic reasoning* and theories of action, yet more new ideas are needed.

learning from experience

The approaches to AI based on *connectionism* and *neural nets* specialize in learning from experience. There is also learning of laws expressed in logic. Programs can only learn what facts or behaviors their formalisms can represent, and unfortunately learning systems are almost all based on very limited abilities to represent information.

planning

Planning programs start with general facts about the world, facts about the particular situation and a statement of a goal. From these, they generate a strategy for achieving the goal. In the most common cases, the strategy is just a sequence of actions.

epistemology

This is a study of the kinds of knowledge that are required for solving problems in the world.

ontology

Ontology is the study of the kinds of things that exist. In AI, the programs and sentences deal with various kinds of objects. Emphasis on ontology began in the 1990s.

heuristics

A heuristic is a way of trying to discover something or an idea imbedded in a program. The term is used variously in AI. *Heuristic functions* are used in some approaches to search to measure how far a node in a search tree seems to be from a goal. *Heuristic predicates* that compare two nodes in a search tree to see if one is better than the other, i.e. constitutes an advance toward the goal, may be more useful.

genetic programming

Genetic programming is a technique for getting programs to solve a task by mating random Lisp programs and selecting fittest in millions of generations.

Functions of AI

There are many functions associated with AI. Some functions can be classified as 'problems' and some as 'techniques'.

* **Automatic Programming :** The task of describing what a program should do and having the AI system 'write' the program.
* **Bayesian Networks :** A technique of structuring and inferencing with probabilistic information. It is a part of the "machine learning" problem.
* **Constraint Satisfaction :** It is solving NP-complete problems, using a variety of techniques.
* **Knowledge Engineering/Representation :** Turning what we know about a particular domain into a form in which a computer can understand it.
* **Machine Learning :** Programs that learn from experience or data.
* **Natural Language Processing(NLP) :** Processing and (perhaps) understanding human ("natural") language. Also known as computational linguistics.
* **Neural Networks (NN) :** The study of programs that function in a manner similar to how animal brains do.
* **Planning :** Given a set of actions, a goal state, and a present state, decide which actions must be taken so that the present state is turned into the goal state.
* **Robotics :** The intersection of AI and robotics, this field tries to get (usually mobile) robots to act intelligently.
* **Speech Recogntion :** Conversion of speech into text.
* **Search :** The finding of a path from a start state to a goal state. Similar to planning, yet different.

* **Visual Pattern Recognition :** The ability to reproduce the human sense of sight on a machine.

Programming Languages for AI

AI programs have been written in just about every language ever created. The most common seem to be Lisp, Prolog, C/C++, recently Java, and even more recently, Python.

LISP : For many years, AI was done as research in universities and laboratories, thus fast prototyping was favored over fast execution. This is one reason why AI has favored high-level langauges such as Lisp. This tradition means that current AI Lisp programmers can draw on many resources from the community. Features of the language that are good for AI programming include: garbage collection, dynamic typing, functions as data, uniform syntax, interactive environment, and extensibility.

PROLOG : This language wins 'cool idea' competition. It wasn't until the 70s that people began to realize that a set of logical statements plus a general theorem prover could make up a program. Prolog combines the high-level and traditional advantages of Lisp with a built-in unifier, which is particularly useful in AI. Prolog seems to be good for problems in which logic is intimately involved, or whose solutions have a succinct logical characterization. Its major drawback (IMHO) is that it's hard to learn.

C/C++ : The speed demon of the bunch, C/C++ is mostly used when the program is simple, and excecution speed is the most important. Statistical AI techniques such as neural networks are common examples of this. Backpropagation is only a couple of pages of C/C++ code, and needs every ounce of speed that the programmer can muster.

Java : The newcomer, Java uses several ideas from Lisp, most notably garbage collection. Its portability makes it desirable for just about any application, and it has a decent set of built in types. Java is still not as high-level as Lisp or Prolog, and not as fast as C, making it best when portability is paramount.

Python : This language does not have widespread acceptance yet, but several people have suggested that it might end up passing Java soon.

AI and Mathematics

AI is a scientific and engineering discipline depending on sophisticated Computer Science techniqes, mathematics, etc. It is also sub-divided into many distinct subfields.

Work in Artificial Intelligence has drawn heavily on that in mathematics in recent years, and a background in mathematics is increasingly important in research. Areas of AI that have particularly benefited from interaction with mathematics include machine learning, neural networks, and advanced logics. The flow has been two way. AI workers have contributed to the automated proof of mathematical theorems (some unproven by humans); to the development of new forms of logic for fuzzy and temporal reasoning; and to the development of new randomised algorithms that have attracted attention from the Mathematics community. In recent years Artificial Intelligence (AI) has become more mathematical.

New uses are being found for mathematics in every area of AI. Mathematics has played some role in AI since its inception. This role is stronger in some areas than others. For instance, mathematical logic has always played a major role in knowledge representation; geometry in visual perception and grammars in natural language understanding. Mathematics has long been a domain of study in AI. Automatic theorem provers were among the first AI programs. Some of the first practical applications of AI research were to the manipulation of mathematical formulae for engineers and scientists.

Applications of AI

Artificial intelligence is an important part of everyday life in developed nations. AI is helping people in every field make better use of information. The AI-development community has generated techniques that are beginning to show promise for solving real business problems involving complex data in dynamic environments - problems such as detecting fraud and automating work flows within and across organizations. The technology isn't appropriate for all information problems — but it does solve some of them very well indeed. AI vendors promote commercial uses of enabling technology. Artificial intelligence is making its way out in the society.

Accomplishments of AI

- Deployed speech dialog systems by firms like IBM, Dragon and Lernout & Hauspie.
- Financial software, which are used by banks to scan credit card transactions for unusual patterns that might signal fraud.
- Applications of expert systems/case-based reasoning for example, a computerized Leukemia diagnosis system did a better job checking for blood disorders than human experts.
- Machine translation for Environment Canada: this software was developed in the 1970s translated natural language weather forcasts between English and French. Purportedly stil in use.
- Physical design analysis programs which are used for buildings and highways.
- Fuzzy controllers in dishwashers, etc.

INTELLIGENT SYSTEMS FOR INFORMATION INFRASTRUCTURE

AI uses the theoretical and experimental tools of computer science to study the phenomena of intelligent behavior. The field addresses a profound scientific problem as well as develops practical technology for constructing intelligent systems. AI research has produced an extensive body of principles, representations, and algorithms. Successful AI applications range from custom-built expert systems to mass-produced software and consumer electronics.

Intelligent Systems are systems that emulate and actively employ some aspect of human intelligence in performing a task. Intelligent Systems that are comprised of sensors, software and computers are embedded in machines basically bringing the power of computing technology into daily lives and business practices. These technologies are enhancing the human ability to perceive, reason, and act. To put this into context, an archetypal intelligent system would allow a machine to autonomously interact with environments that are complex, foreign, hazardous or unpredictable.

Developments in AI

Information and communication technologies are rapidly converging to create machines that understand humans, do what they tell them to, and even anticipate the needs. Scientific advances are making it possible for people to talk to smart computers, while more enterprises are exploiting the commercial potential of the AI. Forecasts conducted under the TechCast Project at George Washington University indicate that 20 commercial aspects of Internet use should reach 30% 'take-off' adoption levels during the second half of this decade to rejuvenate the economy. Meanwhile, the project's technology scanning finds that advances in speech recognition, artificial intelligence, powerful computers, virtual environments, and flat wall monitors are producing a 'conversational' human-machine interface. These powerful trends will drive the next generation of information technology into the mainstream by about 2010. The following are a few of the advances in speech recognition, artificial intelligence, powerful chips, virtual environments, and flat-screen wall monitors that are likely to produce this intelligent interface. IBM has a Super Human Speech Recognition Program to greatly improve accuracy, and in the next decade Microsoft's program is expected to reduce the error rate of speech recognition, matching human capabilities. Project Oxygen of MIT features a voice-machine interface. Amtrak, Wells Fargo, and many other organizations are replacing keypad-menu call centers with speech-recognition systems so that they can improve customer service and recover investment in a year or two. General Motors OnStar driver assistance system uses voice commands, with live staff for backup. The Lexus DVD Navigation System responds to over 100 commands and guides the driver with voice and visual directions. A new technology, called natural language speech recognition, is markedly improving voice-activated self-service. Powered by artificial intelligence, these speech-recognition systems are altering consumer perceptions about phone self-service, as calls for help no longer elicit calls for help. It is proven and moving toward the mainstream.

REFERENCES

1. Abelson, Harold. 1997. Artificial Intelligence and Robotics. In The Computer Science and Engineering Handbook, ed. Tucker Jr, Allen B., 497-499. Boca Raton, FL: CRC Press, Inc.
2. Jarvis, John. F., and Edward Grant. 1997. Intelligent Machines. In McGraw-Hill Encyclopedia of Science and Technology, Volume 9, 309-313. New York: McGraw-Hill.
3. Newell, Allen, and Bruce G. Buchanan.1997. Artificial Intelligence. In McGraw-Hill Encyclopedia of Science and Technology, Volume 2, 146-150. New York: McGraw-Hill.
4. Bobrow, Daniel G., editor. 1994. Artificial Intelligence in Perspective. Cambridge, MA: Bradford Book/MIT Press. Printed first in 1993 in the journal Artificial Intelligence 59: 5-20. Bobrow has brought together thoughts and comments from the authors of several classic papers whose work has been continuously used and referred to by AI scientists. The contributors reflect on the impact their work has had during decades of AI research.
5. Boden, Margaret A. 1977. Artificial Intelligence and Natural Man. New York: Basic Books. Although the examples are somewhat dated, this book remains a clear discussion of the goals and controversial issues of AI.
6. Cawsey, Alison. 1998. The Essence of Artificial Intelligence. Essence of Computing Series. London: Prentice Hall Europe. This book emphasizes practical aspects and main techniques. Link to the author's website for an overview of the book, sample Prolog programs and tutorials.

7. Challoner, Jack. 2002. Artificial Intelligence (part of the Essential Science series). New York: DK Publishing Inc.

8. Fetzer, James H., editor. 1988. Aspects of Artificial Intelligence. Dordrecht, Netherlands: Kluwer Academic Press. A collection of articles by AI professionals. Contributions on the ontological foundations of AI by B. Cantwell Smith, J. H. Moor, J. C. Maloney, W. J. Rapaport, J. H. Fetzer, B. MacLennan. Articles on AI methodology from C. Glymour, B. Buchanan, D. Nute, T. Rankin, K. Kelly, R. A. Vaughan, R. Scheines.

9. Gelernter, David. 1994. The Muse in the Machine: Computerizing the Poetry of Human Thought. New York: The Free Press of Macmillan, Inc.

10. Hogan, James P. 1997. Mind Matters: Exploring the World of Artificial Intelligence. New York: Ballantine Publishing Group. A non-technical look at the development, history, issues and people in the AI field.

11. Kurzweil, Raymond. 1990. The Age of Intelligent Machines. Cambridge, MA: MIT Press.

12. The Age of Intelligent Machines: The Film. By Raymond Kurzweil. "A survey of Artificial Intelligence showing AI at work and under development. The paradoxes, promise and challenges of advanced computer science, with authorities Marvin Minsky, Roger Schank, Raj Reddy and other leaders in the field.

13. McCorduck, Pamela. 2004. Machines Who Think: A Personal Inquiry into the History and Prospects of Artificial Intelligence. A K Peters, Natick, Mass. [Reviewed in Scientific American; May 2004.]

14. Also see Pamela McCorduck's FAQ Collection.

15. Minsky, Marvin. 1985. The Society of Mind. New York: Simon and Schuster.

16. Scientific American. 2002. Understanding Artificial Intelligence. New York: Warner Books, Inc. A collection of essays with a Foreword by Rodney Brooks and an Introduction by Sandy Fritz.

17. Simon, Herbert. 1996. Sciences of the Artificial. 3rd edition. Cambridge, MA: MIT Press. A classic book, originally published in 1969, that examines several presuppositions of AI. Updates throughout the book take into account advances in cognitive science and the science of design.

18. Brooks, R. A. 1991a. Intelligence Without Representation. Artificial Intelligence 47 (1-3): 139-159.

19. Epstein, Robert. 1992. The Quest for the Thinking Computer. AI Magazine 13 (2): 80-95.

20. Feigenbaum, Edward A. 1996. How the 'What' Becomes the 'How'. Communications of the ACM 39 (5): 97-104.

21. "Robot," Microsoft® Encarta® Online Encyclopedia 2006 http://encarta.msn.com/

22. Haag, Stephen. Cummings, Maeve. McCubbrey J, Donald. Pinsonneault, Alain. Donovan, Richard. Management Information Systems for the Information Age. Third Canadian Edition. Canada. McGraw-Hill, 2006.

23. Hillis, Daniel. 1997. Can They Feel Your Pain? Newsweek 129 (May 5, 1997): 57.

24. Lloyd-Williams, M. & Beynon-Davies, P. Expert system for database design: a comparative review. Artificial Intelligence Review, 6, 1992, 263-283.

25. Vossen, G. (1990) Data Models, Database Languages, and Database Management Systems. Wilkinson: Addison-Wesley, 1990.

26. Storey, V.C. A selective survey of the use of artificial intelligence for database design systems. Data and Knowledge Engineering. 11, 1993, 61-102.

27. Bouzeghoub, M., Gardarin, G. & Metais, E. Database design tools: an expert system approach. Proceeding of the 11th International Conference on Very Large Databases, Stockholm, Sweden, 1985, pp. 82-95.

28. Goldstein, R.C. Database: Technology and Management. New York: John Wiley, 1985.

29. *www.gwforecast.gwu.edu*

30. Storey, V.C. & Goldstein, R. Knowledge-based approach to database design. MIS Quarterly. Volume 17, Issue 1, March 1993, 25-46.

31. Lloyd-Williams, M. Knowledge-based CASE tools: improving performance using domain specific knowledge. Software Engineering Journal, Volume 9, Issue 4, 1994, 167-173.

32. Lloyd-Williams, M. & Beynon-Davies, P. Knowledge based CASE tools for database design. In K. Spurr & P. Layzell (Eds.), CASE: Current Practise, Future Prospects. John Wiley, 1992, pp. 205-222.

33. Choobineh, J., Konsynski, B.R., Mannino, M. V. & Nunamaker J. F. An expert database system based on analysis on forms. IEEE Transactions on Software Engineering, 15, 1989, 479-491.

34. Storey, V.C. Real world knowledge for databases. Journal of Database Administration, 3, 1992, 1-19.

35. Lloyd-Williams, M. Expert system support for object-oriented database

design, International Journal of Applied Expert Systems, 1, 1993, 197-212.

36. Concepcion, A.I. & Villafuete, R.M ExpertDB: An assistant database design system. In: Proceeding of the 3rd International Conference on Industrial and Engineering Applications of Artificial Intelligence and Expert Systems, New York, NY: ACM, 1990, pp. 333-340.

37. Ahrens, H.D. A CASE tool knowledge base for semantic data modelling. In Khosrowpour, D.B.A. (Ed.) Challenges for Information Management in a World Economy. Proceedings of 1993 Information Resources Management Association International Conference. Harrisburg PA: Idea Group Publishing, 1993, pp. 81-95.

38. Nengbin, W. & Haiqing, L. An intelligent tool to support requirements analysis and conceptual design of database design. Journal of Computer Science and Technology, 6, 1991, 153-160

39. Rumbaugh, J., Blaha, M., Premerlani, W., Eddy, F. & Lorenson, W. Object-Oriented Modelling and Design. Englewood-Cliffs, NJ: Prentice-Hall, 1991.

40. Baldiserra, C., Ceri, S., Pelagatti, G. & Bracchi, G. Interactive specification and formal verification of user's views in database design. In. Proceeding of the 5th International Conference on Very Large Databases, 1979, pp. 262-272.

41. Atkinson, M., Dewitt, D., Maier, D., Altair, F.B., Dittrich, K. & Zdonik, S. The object-oriented database manifesto. In W. Kim, J.M. Nicholas & S. Nishio (Eds.) Deductive and Object-Oriented Databases. London: Elsevier Science, 1990, pp. 223-240

42. Catell, R.G. Next generation database systems. Communications of the ACM, 34(10), 1991, 30-33.

43. Lenat, Douglas B. 1995. Artificial Intelligence: A Crucial Storehouse of Commonsense Knowledge is Now Taking Shape. Scientific American 273 (3): 80-82.

44. Masci, David. 1997. Artificial Intelligence. (Cover story). Congressional Quarterly Researcher(November 14, 1997): 985. An easy-to-read overview of AI issues, goals, projects, and a short chronology of AI achievements.

45. Newell, Allen. 1983. Some Intellectual Issues in the History of Artificial Intelligence. In The Study of Information: Interdisciplinary Perspectives, ed. Machlup, F. and U. Mansfield, 187-227. New York: Wiley.

46. Newell, Allen, and Herbert Simon. 1976. Computer Science as Empirical Inquiry: Symbols and Search. In Computation and

Intelligence, ed. Luger, George F., 91-119. Menlo Park/Cambridge, MA./London: AAAI Press/The MIT Press, 1995. [Originally published in Communications of the ACM 19(3).]

47. Schank, Roger C. 1990. What is AI, Anyway? In The Foundations of Artificial Intelligence, ed. Partridge, D. and Y. Wilks, 3-13. Cambridge, England: Cambridge University Press.
48. Simon, Herbert 1995. Artificial Intelligence: An Empirical Science. Artificial Intelligence 77 (1): 95-127.
49. Stewart, Doug. Interview with Herbert Simon, June 1994. Omni Magazine. One of the many probing questions is: "What is this the main goal of AI?" [No longer available online.]
50. Ullman, Ellen. 2002. Programming the Post-Human: Computer science redefines "life." Harper's, Vol. 305, No. 1829: 60-70
51. Web references:
52. http://istresults.cordis.lu/index.cfm/section/news/tpl/article/BrowsingType/Features/ID/82453
53. http://www.aaai.org/Organization/organization.php
54. http://www.aaai.org/AITopics/html/faqs.html
55. http://www.gcn.com/vol1_no1/daily-updates/26338-1.html
56. http://www.aaai.org/AITopics/assets/FAQs/AI&Math.html
57. http://www.cfo.com/article.cfm/3515798/c_3516777?f=magazine_alsoinside
58. http://www.businessweek.com/bw50/content/mar2003/a3826072.htm
59. http://www.mckinseyquarterly.com/article_abstract.asp?ar=1201&L2=13&L3=13
60. http://www.vnunet.com/news/1158217
61. http://www.nap.edu/readingroom/books/far/ch9.html
62. http://www.precarn.ca/home/index_en.html
63. http://www.stottlerhenke.com/ai_general/index.htm
64. http://www.aaai.org/Library/Reports/reports-library.php
65. Alan M. Turing. Computing machinery and intelligence. Mind,
66. LIX(236):433-460, October 1950. (*http://www.abelard.org/turpap/turpap.htm*)
67. http://www-formal.stanford.edu/jmc/whatisai/
68. *http://www.paulgraham.com/avg.html*
69. *http://norvig.com/python-lisp.html*

70. http://library.thinkquest.org/2705/
71. http://library.thinkquest.org/2705/applications.html
72. Copeland, Jack —*Artificial intelligence: A Philosophical Introduction* Blackwell Publishers, United Kingdom. copyright (c) 1993
73. Crevier, Daniel —*AI: The Tumultuous History of the Search for Artificial Intelligence* HarperCollins Publishers, Inc. New York, NY. copyright (c) 1993
74. Neural Network basics —*http://alpha.luc.ac.be/~bart/neural.html*
75. Artificial Intelligence: A Modern Approach Stuart Russell, *and* Peter Norvig . Prentice Hall. Copyright: 2003
76. Dubey, Y P. Artificial Intelligence Application in Library and Information Systems. SIS Proceedings, 2003, pp. 479-489.
77. Willaimson, M. Artificial Intellignece for Microcoputers: Guide for Business Decision Makers. New York: Brady Communications, 1986.
78. Barr, A and Feigenbaum, L A (Ed.). The Handbook of Artificial Intelligence. Reading, M A: Addison Wesley, 1987.
79. http://www.wtec.org/loyola/kb/c2_s2.htm

14
Data Mining

Generally, data mining (sometimes called data or knowledge discovery) is the process of analyzing data from different perspectives and summarizing it into useful information - information that can be used to increase revenue, cuts costs, or both. Data mining software is one of a number of analytical tools for analyzing data. It allows users to analyze data from many different dimensions or angles, categorize it, and summarize the relationships identified. Technically, data mining is the process of finding correlations or patterns among dozens of fields in large relational databases. Although data mining is a relatively new term, the technology is not. Companies have used powerful computers to sift through volumes of supermarket scanner data and analyze market research reports for years. Continuous innovations in computer processing power, disk storage, and statistical software are dramatically increasing the accuracy of analysis.

Data mining has been defined as "The nontrivial extraction of implicit, previously unknown, and potentially useful information from data". Data mining uses machine learning, statistical and visualization techniques to discovery and present knowledge in a form. This knowledge is easily comprehensible to humans. Data Mining can also be defined as "The nontrivial extraction of implicit, previously unknown, and potentially useful information from data" and "The science of extracting useful information from large data sets or databases". Although it is usually used in relation to analysis of data, data mining, like artificial intelligence, is an umbrella term and can be used in a wide range of contexts. It is usually associated with identifing trends. Data mining involves the process of analysing data to show patterns or relationships; sorting through large amounts of data; and picking out pieces of relative information or patterns that occur.

Data mining, *the extraction of hidden predictive information from large databases*, is a powerful new technology with great potential to help organisations focus on the most important information in their data warehouses. Data mining tools predict future trends and behaviors, allowing businesses to make proactive, knowledge-driven decisions. Data mining tools can answer business questions that traditionally were too time consuming to resolve. Data mining tools scour identify hidden patterns and find predictive information.

NEED OF DATA MINING:

The goals of data mining in practice are prediction and description. Prediction involves using some variables or fields in the database to predict unknown or future values of other variables of interest, and description focuses on finding human-interpretable patterns describing the data. The goals of prediction and description can be achieved using a variety of particular data-mining methods like,

* Classification
* Clustering
* Association Rules
* Sequential Patterns
* Regression
* Summarization
* Change and Deviation Detection

Most organisations already collect and refine massive

quantities of data. The important feature is that data mining techniques can be implemented rapidly on existing software and hardware platforms to enhance the value of existing information resources, and these tecniques can be integrated with new products and systems as they are brought on-line. Data mining tools can analyze massive databases to deliver answers to questions when implemented on high performance client/server or parallel processing computers.

Essentials of Data Mining

Data mining techniques are the result of a long process of research and product development. Data mining takes the evolutionary process beyond retrospective data access and navigation to prospective and proactive information delivery. Data mining is ready for application in the organisations due to by three technologies that are given below :

* Massive data collection
* Powerful multiprocessor computers
* Data mining algorithms

HOW DOES DATA MINING WORK?

Data mining provides link between separate transaction and analytical systems. Data mining software analyzes relationships and patterns in stored transaction data which is based on open-ended user queries. Several types of analytical software are available such as statistical, machine learning, and neural networks. Generally, any of four types of relationships are sought:

* **Classes** : Stored data is used to locate data in predetermined groups.
* **Clusters** : Data items are grouped according to logical relationships or consumer preferences.
* **Associations** : Data can be mined to identify associations.
* **Sequential patterns** : Data is mined to anticipate behavior patterns and trends.

ELEMENTS OF DATA MINING

Data mining consists of five major elements:

* Extract, transform, and load transaction data onto the data warehouse system.
* Store and manage the data in a multidimensional database system.
* Provide data access to business analysts and information technology professionals.
* Analyze the data by application software.
* Present the data in a useful format, such as a graph or table.

TECHNIQUES IN DATA MINING

The most commonly used techniques in data mining and different levels of analysis are as following:

Artificial neural networks : These are non-linear predictive models that learn through training and resemble biological neural networks in structure.

Genetic algorithms : Optimization techniques that use process such as genetic combination, mutation, and natural selection in a design based on the concepts of natural evolution.

Decision trees : These are tree-shaped structures that represent sets of decisions. These decisions generate rules for the classification of a dataset. Specific decision tree methods include Classification and Regression Trees (CART) and Chi Square Automatic Interaction Detection (CHAID). They provide a set of rules that user can apply to a new (unclassified) dataset to predict which records will have a given outcome.

Nearest neighbor method : This technique classifies each record in a dataset based on a combination of the classes of the *k* record(s) most similar to it in a historical dataset (where *k* 1). Sometimes it is called the *k*-nearest neighbor technique.

Rule induction : The extraction of useful if-then rules from data based on statistical significance.

Data visualization : The visual interpretation of complex relationships in multidimensional data. Graphics tools are used to illustrate data relationships.

USING DATA MINING TOOLS

Data mining software are analysed into two groups: data mining tools and data mining applications. Data mining tools provide a number of techniques that can be applied to any business problem. Data mining applications, on the other hand, embed techniques inside an application customized to address a specific business problem. Both data mining tools and data mining applications are valuable. Organizations are using data mining tools and data mining applications together in an integrated environment for predictive analytics. Data mining tools are used to ensure flexibility and the greatest accuracy possible. It is found that, data mining tools increase the effectiveness of data mining applications. Data mining tools deliver in-depth techniques as well as flexibility to use combinations of techniques to improve predictive accuracy.

THE SCOPE OF DATA MINING

Data mining derives its name from the similarities between searching for valuable business information in a large database — It requires either sifting through an immense amount of material, or intelligently probing it to find exactly where the value resides. Scope of data mining is as following :

Automated prediction of trends and behaviors : Data mining automates the process of finding predictive information in large databases. Extensive hands-on analysis can now be answered directly from the data — quickly. For example, in targeted marketing, it uses data on past promotional mailings to identify the targets most likely to maximize return on investment in future mailings.

Automated discovery of previously unknown patterns: Data mining tools sweep through databases and identify previously hidden patterns in one step. The pattern discovery problems include detecting fraudulent credit card transactions and identifying anomalous data that could represent data entry keying errors.

Importance:

Data mining techniques can yield the benefits of automation on existing software and hardware platforms, and these can be implemented on new systems as existing platforms are upgraded and new products developed. Due to being implemented on high performance parallel processing systems, data mining tools can analyze massive databases in minutes. Faster processing means that users can automatically experiment with more models to understand complex data. High speed makes it practical for users to analyze huge quantities of data. Larger databases, in turn, yield improved predictions. It is further explained as :

1. ***More columns*** : Analysts must often limit the number of variables they examine when doing hands-on analysis due to time constraints. Though variables that are discarded because they seem unimportant may carry information about unknown patterns.

2. ***More rows*** : The larger samples yield lower estimation errors and variance, and they also allow users to make inferences about small but important segments of a population.

It is found that data mining uncovers patterns in data using predictive techniques. These patterns play a critical role in decision making because they reveal areas for process improvement. Using data mining, organizations can increase the profitability of their interactions with customers, detect fraud, and improve risk management.

Architecture for Data Mining

Many data mining tools currently operate outside of the warehouse, requiring extra steps for extracting, importing, and analyzing the data. Furthermore, when new insights require operational implementation, integration with the warehouse simplifies the application of results from data mining. The resulting analytic data warehouse can be applied to improve business processes throughout the organization, in areas such as promotional campaign management, fraud detection, new product rollout, and so on. Following figure illustrates architecture for advanced analysis in a large data warehouse.

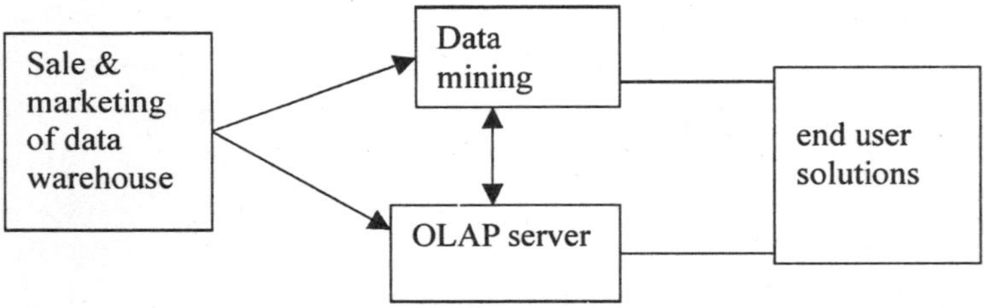

Fig. 1. Integrated Data Mining Architecture

The ideal starting point is a data warehouse containing a combination of internal data tracking all customer contact coupled with external market data about competitor activity. Background information on potential customers also provides an excellent basis for prospecting. This warehouse can be implemented in a variety of relational database systems: Sybase, Oracle, Redbrick, and so on, and should be optimized for flexible and fast data access.

An OLAP (On-Line Analytical Processing) server enables a more sophisticated end-user business model to be applied when navigating the data warehouse. The multidimensional structures allow the user to analyze the data as they want to view their business – summarizing by product line, region, and other key perspectives of their business. The Data Mining Server must be integrated with the data warehouse and the OLAP server to embed ROI-focused (Region of Interest) business analysis directly into this infrastructure. An advanced, process-centric metadata template defines the data mining objectives for specific business issues like campaign management, prospecting, and promotion optimization. Integration with the data warehouse enables operational decisions to be directly implemented and tracked. As the warehouse grows with new decisions and results, the organization can continually mine the best practices and apply them to future decisions.

This design represents a fundamental shift from conventional decision support systems. Rather than simply delivering data to the end user through query and reporting software, the Advanced Analysis Server applies users' business models directly to the warehouse and returns a proactive analysis of the most relevant information. These results enhance the metadata in the OLAP Server by providing a dynamic metadata layer that represents a distilled view of the data. Reporting, visualization, and other analysis tools can then be applied to plan future actions and confirm the impact of those plans.

DATA MINING AND CLASSIFICATION

Data mining creates classification models by examining already classified data (cases) and inductively finding a predictive pattern. These existing cases may come from an historical database, such as people who have already undergone a particular medical treatment or moved to a new long distance service. They may come from an experiment in which a sample of the entire database is tested in the real world and the results used to create a classifier. Sometimes an expert classifies a sample of the database, and this classification is then used to create the model which will be applied to the entire database.

Classification, perhaps the most commonly applied data mining technique, employs a set of pre-classified examples to develop a model that can classify the population of records at large. Fraud detection and credit-risk applications are particularly well suited to this type of analysis. This approach frequently employs decision tree or neural network-based classification algorithms. The use of classification algorithms begins with a training set of pre-classified example transactions. For a fraud detection application, this would include complete records of both fraudulent and valid activities, determined on a record-by-record basis. The classifier training algorithm uses these pre-classified examples to determine the set of parameters required for proper discrimination. The algorithm then encodes these parameters into a model called a classifier. The approach affects the explanation capability of the system. Once an effective classifier is developed, it is used in a predictive mode to classify new records into these same predefined classes.

DATA MINING AND CLUSTERING

Clustering approaches address segmentation problems. These approaches assign records with a large number of attributes into a relatively small set of groups or "segments." This assignment process is performed automatically by clustering algorithms that identify the distinguishing characteristics of the dataset and then partition the n-dimensional space defined by the dataset attributes along natural cleaving boundaries. There is no need to identify the groupings desired or the attributes that should be used to segment the dataset.

Clustering is often one of the first steps in data mining analysis. It identifies groups of related records that can be used as a starting point for exploring further relationships. This technique supports the development of population segmentation models, such as demographic-based customer segmentation. Additional analyses using standard analytical and other data mining techniques can determine the characteristics of these segments with respect to some desired outcome. For example, the buying habits of multiple population segments might be compared to determine which segments to target for a new sales campaign.

Clustering divides a database into different groups. The goal of clustering is to find groups that are very different from each other, and whose members are very similar to each other. One should not confuse clustering with segmentation. Segmentation refers to the general problem of identifying groups that have common characteristics. Clustering is a way to segment data into groups that are not previously defined, whereas classification is a way to segment data by assigning it to groups that are already defined.

* Application areas
* Business intelligence
* Business performance management
* Discovery Science
* Loyalty card
* Cheminformatics
* Quantitative structure-activity relationship
* Bioinformatics

WORKING OF DATA MINING

How exactly is data mining able to tell one important thing that someone didn't know or what is going to happen next? The technique that is used to perform these feats in data mining is called modeling. Modeling is simply the act of building a model in one situation where someone knows the answer and then applying it to another situation that someone don't. Hopefully, if someone has got a good model, he finds his treasure. This act of model building is thus something that people have been doing for a long time, certainly before the advent of computers or data mining technology. What happens on computers, however, is not much different than the way people build models. Computers are loaded up with lots of information about a variety of situations where an answer is known and then the data mining software on the computer must run through that data and distill the characteristics of the data that should go into the model. Once the model is built it can then be used in similar situations where one doesn't know the answer.

ADVANCES IN DATA MINING

Recent advances have led to the newest and hottest trends

in data mining—text mining and Web mining. These two data mining technologies open a rich vein of customer data in the form of textual comments from survey research and log files from Web servers, which were previously unusable. Applying data mining to these data adds a richness and depth to the patterns already uncovered through the data mining efforts.

Text mining

Text mining enables organizations to explore the "unstructured" information contained in text in much the same way that data mining explores tabular or "structured" data. Through text mining, one can uncover hidden patterns, relationships, and trends in text. As a result, he gains greater insight from articles, reports, surveys, call center notes fields, e-mail, RSS (Rich Site Summary) feeds—such as blogs and news feeds, and other types of text documents.

In addition, sodtware enables to build on the text mining efforts with products that help to incorporate text in predictive data mining models—a solution that is called Predictive Text Analytics. Predictive Text Analytics provides a way for the organization to combine structured and unstructured information in the same models. This enables to draw more reliable conclusions and take more effective action.

By using Predictive Text Analytics, the organization can:

* Uncover concepts and relationships among concepts in text that would be too costly—or even impossible—to detect with other methods
* Improve the "lift" or accuracy of predictive models and, by doing so, make the models more effective
* Deploy results both to the people who make decisions and to automated systems that make recommendations

Web mining

Traditional Web analytics are often failing to provide the actionable information decision makers need to successfully manage their online business. Predictive Web Analytics solutions help companies close the online insight gaps by combining the leading technologies for predictive analytics and enterprise Web analytics—along with software more than 35 years of analytical experience. Predictive Web Analytics solutions deliver predictive analytic application modules for online insight through an enterprise architecure. These application modules address critical online business goals like increasing conversion rates with predictive analysis capabilities that range from automated visitor segmentation to prescriptive search engine marketing intelligence. Predictive intelligence requires little interpretation and can be used to immediately improve online customer interactions. For example, quantifying "likelihood to convert by individual visitor" takes users far beyond "number of visits" by providing an actual list of the visitors most likely to convert. This online insight is both predictive in that it helps organizations to anticipate customer needs and prescriptive in delivering plain-English recommendations on how to meet them.

REFERENCES

1. W. Frawley and G. Piatetsky-Shapiro and C. Matheus, Knowledge Discovery in Databases: An Overview. AI Magazine, Fall 1992, pp. 213-228.
2. D. Hand, H. Mannila, P. Smyth: Principles of Data Mining. MIT Press, Cambridge, MA, 2001.
3. A.G. Ivakhnenko, Heuristic Self-Organization in Problems of Engineering Cybernetics, GMDH library, Automatica, 6, 1970, pp.207–219.
4. Fred Schwed, Jr, Where Are the Customers' Yachts? ISBN 0471119792 (1940).
5. T. Menzies, Y. Hu, Data Mining For Very Busy People. IEEE Computer, October 2003, pp. 18-25.
6. K.A. Taipale, Data Mining and Domestic Security: Connecting the Dots to Make Sense of Data, Center for Advanced Studies in Science and Technology Policy.
7. http://www.spss.com/data_mining/
8. www.the-data-mine.com
9. www.kdnuggets.com
10. Caudill, M. & Butler, C. (1990) *Naturally Intelligent Systems*. Cambridge: MIT Press.

11. Allee, V. (2000). Reconfiguring the value network, Journal of Business Strategy, **21**(4), 36-39.
12. Boudreau, G. and Manley, C. (2004). Developing a customized database of users as a tool for marketing the library, Information Outlook, **8**(1), 44.
13. Buren, A., Schierholz, R., Kolbe, L., and Brenner, W. (2004). Customer knowledge management - improving performance of customer relationship management with knowledge management, Proceedings of the 37th Hawaii Conference on System Science.
14. Chelton, M. K. (2001). When Oprah meets email: virtual book clubs, Reference & User Services Quarterly, **41**(1), 31-36.
15. Chelton, M. K. (2003). Readers advisory 101: a crash course in RA: common mistakes librarians make and how to avoid them, Library Journal, 11/1/2003. http://www.libraryjournal.com/article/CA329318.html.
16. Cook, J. (2006). Shelfari an online meeting place for bibliophiles, Seattle Post-Intelligencer, October 11, 2006. http://seattlepi.nwsource.com/business/288229_shelfari11.html.
17. Desouza, K. C. and Awazu, Y. (2005). What do they know? (customer knowledge management), Business Strategy Review, **16**(1), 41-45.
18. Fayyad, U. & Uthurasamy, R. (1996) "Data Mining and Knowledge Discovery in Databases" *Communications of the ACM*, 39(11), 24-26.
19. Fayyad, U., Piatetsky-Shapiro, G. & Smyth, P. (1996) "The KDD Process for Extracting Useful Knowledge from Volumes of Data" *Communications of the ACM*, 39(11), 27-34.
20. Fayyad, U., Piatetsky-Shapiro, G., Smyth, P. & Uthurasamy, R. (1996) *Advances in Knowledge Discovery and Data Mining*. Cambridge: AAAI/MIT Press.
21. Hedberg, S.R. (1995) "The Data Gold Rush" *Byte*, 20(10), 83-88.
22. Kohonen, T. (1988) *Self-Organisation and Associative Memory*. New York: Springer-Verlag.
23. Limb, P.R., & Meggs, G.J. (1995) "Data Mining -Tools and Techniques" *British Telecom Technology Journal*, 12(4), 32-41.
24. Lloyd-Williams, M., Jenkins, J., Howden-Leach, H., Mathur, M., Morris, C & Cooke, I. (1995) "Knowledge Discovery in an Infertility Database Using Artificial Neural Networks" *IEE Colloquium on Knowledge Discovery in Databases*. IEE Digest No:1995/021(B).
25. Lloyd-Williams, M. & Williams, S. (1996) "A Neural Network Approach to Analysing Healthcare Information", *Topics in Health Information Management*, 17(2), 26-33.
26. Lloyd-Williams, M., Williams, S, Bath, P. & Morris, C. (1996)

"Knowledge Discovery in the WHO *Health for All* Database: Developing A Taxonomy of Mortality Patterns for European Countries", In: Richards, B. & de Glanville,H. (eds) *Current Perspectives in Healthcare Computing. Proceedings of HC96,* pp. 551-556. Weybridge: BJHC.

27. Murtagh, F. & Hernández-Pajares, M. (1995) "The Kohonen Self-Organizing Map Method: An Assessment" *Journal of Classification* 12, 165-190.
28. Scarfe, R. & Shortland, R.J. (1995) "Data Mining Applications in BT" *IEE Colloquium on Knowledge Discovery in Databases.* IEE Digest No:1995/021(B).
29. SPSS Inc. (1995) *Neural Connection Applications Guide.* Chicago: SPSS Inc.
30. Watterson, K. (1995) "A Data Miner's Tools" *Byte* 20(10), 91-96. 4.
31. Wilkins, M. F., Boddy, L., & Morris, C. W. (1993) "Kohonen Maps and Learning Vector Quantization: Neural Networks for Analysis of Multivariate Biological Data" *Binary*, 6, 64-72.
32. www.thearling.com/text/dmwhite/dmwhite.htm
33. http://databases.about.com/od/datamining/a/datamining.htm
34. http://databases.about.com/od/datamining/Data_Mining_and_Data_Warehousing.htm
35. http://databases.about.com/od/datamining/g/clustering.htm
36. META Group Application Development Strategies: "Data Mining for Data Warehouses: Uncovering Hidden Patterns."
37. Gartner Group Advanced Technologies and Applications Research Note.

15
Expert Systems

Expert system is the area of AI that can claim a large measure of responsibility for the current heightened AI awareness. It is the area responsible for developing computer programs that embody human expertise. As one of the first AI technologies to help people solve important problems, the expert system field has become imbued with a certain mystique, a high tech aura that illuminates the field and at the same time, discourages closer examination. Expert systems hold much promise, but they also have been subjected to a great deal of 'hype' as they have moved into the commercial market.

While describing an expert system, it can be said that expert system is a computer program that contains some of the subject-specific knowledge of one or more human experts. The initial developments took place during 1960s and 1970s and were applied commercially throughout the 1980s. The most common form of expert systems is a program made up of a set of rules that analyze information (usually supplied by the user of the system) about a specific class of problems, as well as providing analysis of the problem(s), and, *depending upon their design*, recommend a course of user action in order to implement corrections. In spite of the variety of problems addressed in artificial intelligence research, a number of important features emerge that seem common to all divisions of the field; these include:

1. The use of computers to do reasoning, pattern recognition, learning, or some other form of inference.
2. A focus on problems that do not respond to algorithmic solutions. This underlies the reliance on heuristic search as an AI problem-solving technique.
3. A concern with problem solving using inexact, missing, or poorly defined information and the use of representational formalisms that enable the programmer to compensate for these problems.
4. Reasoning about the significant qualitative features of a situation.
5. An attempt to deal with issues of semantic meaning as well as syntactic form.
6. Answers that are neither exact nor optimal, but are in some sense "sufficient". This is a result of the essential reliance on heuristic problem-solving methods in situations where optimal or exact results are either too expensive or not possible.
7. The use of large amounts of domain-specific knowledge in solving problems. This is the basis of expert systems.
8. The use of meta-level knowledge to effect more sophisticated control of problem solving strategies. Although this is a very difficult problem, addressed in relatively few current systems, it is emerging as an essential area of research.

According to Encyclopædia Britannica, "The basic components of an expert system are a knowledge base, or KB, and an inference engine. The information to be stored in the KB is obtained by interviewing people who are expert in the area in question. The interviewer, or knowledge engineer, organizes the information elicited from the experts into a collection of rules, typically of an 'if-then' structure. Rules of this type are called production rules. The inference engine enables the expert system to draw deductions from the rules in the KB.

Expert systems are computer programs that are derived from a branch of computer science research called *Artificial Intelligence* (AI). AI's scientific goal is to understand intelligence by building computer programs that exhibit intelligent behavior. It is concerned with the concepts and methods of symbolic inference, or reasoning, by a computer, and how the knowledge used to make those inferences will be represented inside the machine. Of course, the term *intelligence* covers many cognitive skills, including the ability to solve problems, learn, and understand language; AI addresses all of those. But most progress to date in AI has been made in the area of problem solving- concepts and methods for building programs that *reason* about problems rather than calculate a solution.

AI programs that achieve expert-level competence in solving problems in task areas by bringing to bear a body of knowledge about specific tasks are called *knowledge-based* or *expert systems*. Often, the term expert systems is reserved for programs whose knowledge base contains the knowledge used by human experts, in contrast to knowledge gathered from textbooks or non-experts. More often than not, the two terms, expert systems (ES) and knowledge-based systems (KBS) are used synonymously. Taken together, they represent the most widespread type of AI application. The area of human intellectual endeavor to be captured in an expert system is called the *task domain*. *Task* refers to some goal-oriented, problem-solving activity. *Domain* refers to the area within which the task is being performed. Typical tasks are diagnosis, planning, scheduling, configuration and design. Building an expert system is known as *knowledge engineering* and its practitioners are called *knowledge engineers*. The knowledge engineer must make sure that the computer has all the knowledge needed to solve a problem. The knowledge engineer must choose one or more forms in which to represent the required knowledge as symbol patterns in the memory of the computer — that is, he (or she) must choose a *knowledge representation*. He must also ensure that the computer can use the knowledge efficiently by selecting from a handful of *reasoning methods*.

THE BUILDING BLOCKS OF EXPERT SYSTEMS

Every expert system consists of two principal parts: the knowledge base; and the reasoning, or inference, engine.

The *knowledge base* of expert systems contains both factual and heuristic knowledge. *Factual knowledge* is that knowledge of the task domain that is widely shared, typically found in textbooks or journals, and commonly agreed upon by those knowledgeable in the particular field.

Heuristic knowledge is the less rigorous, more experiential, more judgmental knowledge of performance. In contrast to factual knowledge, heuristic knowledge is rarely discussed, and is largely individualistic. It is the knowledge of good practice, good judgment, and plausible reasoning in the field. It is the knowledge that underlies the "art of good guessing."

Knowledge representation formalizes and organizes the knowledge. One widely used representation is the *production rule*, or simply *rule*. A rule consists of an IF part and a THEN part (also called a *condition* and an *action*). The 'IF' part lists a set of conditions in some logical combination. The piece of knowledge represented by the production rule is relevant to the line of reasoning being developed if the IF part of the rule is satisfied; consequently, the THEN part can be concluded, or its problem-solving action taken. Expert systems whose knowledge is represented in rule form are called *rule-based systems*.

Another widely used representation, called the *unit* (also known as *frame*, *schema*, or *list structure*) is based upon a more passive view of knowledge. The unit is an assemblage of associated symbolic knowledge about an entity to be represented. Typically, a unit consists of a list of properties of the entity and associated values for those properties.

Since every task domain consists of many entities that stand in various relations, the properties can also be used to specify relations, and the values of these properties are the names of other units that are linked according to the relations. One unit can also represent knowledge that is a "special case" of another unit, or some units can be "parts of" another unit.

The *problem-solving model,* or *paradigm*, organizes and controls the steps taken to solve the problem. One common but powerful paradigm involves chaining of IF-THEN rules to form a line of reasoning. If the chaining starts from a set of conditions and moves toward some conclusion, the method is called *forward chaining*. If the conclusion is known (for example, a goal to be achieved) but the path to that conclusion is not known, then reasoning backwards is called for, and the method is *backward chaining*. These problem-solving methods are built into program modules called *inference engines* or *inference procedures* that manipulate and use knowledge in the knowledge base to form a line of reasoning.

The *knowledge base* an expert uses is what he learned from years of experience. Presumably the more experience he has, the larger his store of knowledge. Knowledge allows him to interpret the information in his databases to advantage in diagnosis, design, and analysis.

Though an expert system consists primarily of a knowledge base and an inference engine, a couple of other features are worth mentioning: reasoning with uncertainty, and explanation of the line of reasoning.

Knowledge is almost always incomplete and uncertain. To deal with uncertain knowledge, a rule may have associated with it a *confidence factor* or a weight. The set of methods for using uncertain knowledge in combination with uncertain data in the reasoning process is called *reasoning with uncertainty*. An important subclass of methods for reasoning with uncertainty is called "fuzzy logic," and the systems that use them are known as "fuzzy systems." Because an expert system uses uncertain or heuristic knowledge its credibility is often in question like humans. Most expert systems have the ability to answer questions of the form: "Why is the answer X?" Explanations can be generated by tracing the line of reasoning used by the inference engine.

The most important ingredient in any expert system is knowledge. The power of expert systems resides in the specific, high-quality knowledge they contain about task domains. AI researchers will continue to explore and add to the current repertoire of knowledge representation and reasoning methods. But in knowledge resides the power. Because of the importance of knowledge in expert systems and because the current knowledge acquisition method is slow and tedious, much of the future of expert systems depends on breaking the knowledge acquisition bottleneck and in codifying and representing a large knowledge infrastructure.

Functional Components

What the system does (rather than how)

"... a computer program that behaves like a human expert in some useful ways."

* **Problem area**
  - solve problems efficiently and effectively in a narrow problem area.
  - typically, pertains to problems that can be symbolically represented.
* **Problem difficulty**
  - apply expert knowledge to difficult real world problems.
  - solve problems that are difficult enough to require significant human expertise for their solution.
  - address problems normally thought to require human specialists for their solution.
* **Performance requirement**
  - the ability to perform at the level of an expert.
  - programs that mimic the advice-giving capabilities of human experts.

— matches a competent level of human expertise in a particular field.

— can offer intelligent advice or make an intelligent decision about a processing function.

— allows a user to access this expertise in a way similar to that in which he might consult a human expert, with a similar result.

* **Explain reasoning**

  — the capability of the system, on demand, to justify its own line of reasoning in a manner directly intelligible to the enquirer.

  — incorporation of explanation processes.

Structural Components

The structural components explain how the expert system functions.

Various components and concerned functions are explained as following :

* **Use AI techniques**

  — using the programming techniques of artificial intelligence, especially those techniques developed for problem solving.

* **Knowledge component**

  — the embodiment within a computer of a knowledge-based component, from an expert skil.

  — a computer based system in which representations of expertise are stored.

  — Expert systems are sophisticated computer programs that manipulate knowledge to solve problems.

* **Separate knowledge and control**

  — make domain knowledge explicit and separate from the rest of the system.

* **Use inference procedures - heuristics - uncertainty**

  — an intelligent computer program that uses knowledge

and inference procedures.

— The style adopted to attain these characteristics is rule-based programming.

— Exhibit intelligent behavior by skillful application of heuristics.

— The 'heuristics' are mostly private, little rules of good judgment (rules of plausible reasoning, rules of good guessing) that characterize expert-level decision making in the field.

— incorporation of ways of handling uncertainty.

* **Model human expert**

— can be thought of as a model of the expertise of the best practitioners of the field.

— representation of domain-specific knowledge in the manner in which the expert thinks.

— involving the use of appropriate information acquired previously from human experts.

KNOWLEDGE ENGINEERING: BASIS OF EXPERT SYSTEMS

Expert systems as they exist today are made of three parts — a base of task-specific knowledge, a set of rules for making decisions about that knowledge, and a means of answering people's questions about the reasons for the program's recommendations. The 'expert' program chooses among alternatives using some of the same rules of thumb that human experts use. In the 1980s, there was little question that expert systems can be highly effective, if not superior to human expertise, in certain highly specialized fields. Expert systems are now in commercial and research use in a number of fields.

Knowledge engineering is the art of designing and building expert systems, and knowledge engineers are its practitioners. Knowledge engineering is the same, perhaps more so. Knowledge engineering is an applied part of the science of artificial intelligence which, in turn, is a part of computer science. Theoretically, then, a knowledge engineer is a computer scientist who knows how to design and implement programs that incorporate artificial intelligence techniques. The nature of knowledge engineering is changing, however, and a new breed of knowledge engineers is emerging. The Role of the Knowledge Engineer is described as following :

* **Knowledge acquisition**
  — the process of acquiring the knowledge from human experts or other sources.
  — also involve developing knowledge to solve the problem
* **knowledge elicitation**
  — coaxing information out of human experts
* **Knowledge representation**
  — Method used to encode the knowledge for use by the expert system
  — Common knowledge representation methods include rules, frames, and cases.
  — Putting the knowledge into rules or cases or patterns is the knowledge representation process

Today there are two ways to build an expert system. They can be built from scratch, or built using a piece of development software known as a "tool" or a "shell." Though different styles and methods of knowledge engineering exist, the basic approach is the same: a knowledge engineer interviews and observes a human expert or a group of experts and learns what the experts know, and how they reason with their knowledge. The engineer then translates the knowledge into a computer-usable language, and designs an inference engine, a reasoning structure, that uses the knowledge appropriately. He also determines how to integrate the use of uncertain knowledge in the reasoning process, and what kinds of explanation would be useful to the end user.

Next, the inference engine and facilities for representing

knowledge and for explaining are programmed, and the domain knowledge is entered into the program piece by piece. It may be that the inference engine is not just right; the form of knowledge representation is awkward for the kind of knowledge needed for the task; and the expert might decide the pieces of knowledge are wrong. All these are discovered and modified as the expert system gradually gains competence. The discovery and cumulation of techniques of machine reasoning and knowledge representation is generally the work of artificial intelligence research. The discovery and cumulation of knowledge of a task domain is the province of domain experts. Domain knowledge consists of both formal, textbook knowledge, and experiential knowledge - the *expertise* of the experts.

TOOLS, SHELLS, AND SKELETONS OF EXPERT SYSTEM

Compared to the wide variation in domain knowledge, only a small number of AI methods are known that are useful in expert systems. Thus, systems can be built that contain these useful methods without any domain-specific knowledge. Such systems are known as *skeletal systems, shells*, or simply *AI tools*. Building expert systems by using shells offers significant advantages such as a system can be built to perform a unique task by entering into a shell all the necessary knowledge about a task domain. The inference engine that applies the knowledge to the task at hand is built into the shell. If the program is not very complicated and if an expert has had some training in the use of a shell, the expert can enter the knowledge himself. The currently available shells range from general-purpose shells to shells custom-tailored to a class of tasks, such as financial planning or real-time process control. In general shells don't help with knowledge acquisition, the task of endowing expert systems with knowledge, a task currently performed by knowledge engineers. The choice of reasoning method, or a shell. The power of an expert system lies in its store of knowledge about the task domain — the more knowledge a system is given, the more competent it becomes.

Expert systems are constructed by obtaining the knowledge from a human expert and coding it into a form that a computer may apply to similar problems. This reliance on the knowledge of a human domain expert for the system's problem solving strategies is a major feature of expert systems. The domain expert provides the necessary knowledge of the problem domain through a general discussion of her problem-solving methods and by demonstrating those skills on a carefully chosen set of sample problems. The AI specialist, or knowledge engineer, as expert systems designers are often known, is responsible for implementing this knowledge in a program that is both effective and seemingly intelligent in its behavior. Once such a program has been written, it is necessary to refine its expertise through a process of giving it example problems to solve, letting the domain expert criticize its behavior, and making any required changes or modifications to the program's knowledge. This process is repeated until the program has achieved the desired level of performance. One of the earliest systems to exploit domain-specific knowledge in problem solving was DENDRAL, developed at Stanford in the late 1960s. DENDRAL was designed to infer the structure of organic molecules from their chemical formulas and mass spectrographic information about the chemical bonds present in the molecules. DENDRAL's methods proved remarkably effective, routinely finding the correct structure out of millions of possibilities after only a few trials. The approach has proved so successful that descendants of the system are used in chemical and pharmaceutical laboratories throughout the world. Whereas DENDRAL was one of the first programs to effectively use domain-specific knowledge to achieve expert level problem-solving performance, MYCIN established the methodology of contemporary expert systems. MYCIN uses expert medical knowledge to diagnose and prescribe treatment for spinal meningitis and bacterial infections of the blood. MYCIN provided clear and logical explanations of its reasoning, used a control structure appropriate to the specific problem domain, and identified criteria to reliably evaluate its performance. Other classic expert systems include the PROSPECTOR program for determining the probable location and type of ore deposits based on geological information about a site, the INTERNIST program for performing diagnosis in the area of internal medicine, the Dipmeter Advisor for interpreting the results of oil well drilling logs, and XCON for configuring VAX computers. XCON was developed in 1981, and at one time every VAX sold by Digital Equipment Corporation was configured by that software. Numerous other expert systems are currently solving problems in areas such as medicine, education, business, design, and science.

VARIOUS EXPERT SYSTEMS

* **ART :** An early general-purpose programming language used in the development of expert systems
* **CLIPS :** Programming language used in the development of expert systems
* **CADUCEUS (expert system) :** Blood-borne infectious bacteria
* **Dendral :** Analysis of mass spectra
* **Dipmeter Advisor :** Analysis of data gathered during oil exploration
* **Informavores Firefly :** Browser based expert systems environment built using flow charts and used in the development of expert systems
* **Jess :** Java Expert System Shell. A CLIPS engine implemented in Java used in the development of expert systems
* **Mycin :** Diagnose infectious blood diseases and recommend antibiotics (by Stanford University)
* **NEXPERT Object :** An early general-purpose commercial backwards-chaining inference engine used in

the development of expert systems

* **Prolog :** Programming language used in the development of expert systems
* **R1 (expert system)/XCon :** Order processing
* **SHINE Real-time Expert System :** Spacecraft Health INference Engine
* **STD Wizard :** Expert system for recommending medical screening tests
* **PyKe :** Pyke is a knowledge-based inference engine (expert system) http://pyke.sourceforge.net/)

It is interesting to note that most expert systems have been written for relatively specialized, expert level domains. These domains are generally well studied and have clearly defined problem-solving strategies. In spite of the promise of expert systems, it would be a mistake to overestimate the ability of this technology. Current deficiencies include:

1. Difficulty in capturing "deep" knowledge of the problem domain.
2. Lack of robustness and flexibility.
3. Inability to provide deep explanations.
4. Difficulties in verification. This is a serious problem, as expert systems technology is being applied to critical applications such as air traffic control, nuclear reactor operations, and weapons systems.
5. Little learning from experience. Current expert systems are handcrafted; once the system is completed, its performance will not improve without further attention from its programmers, leading to doubts about the intelligence of such systems.

THE APPLICATIONS OF EXPERT SYSTEMS

The spectrum of applications of expert systems technology to industrial and commercial problems is very wide. The applications find their way into most areas of knowledge work. Applications tend to cluster into following major classes.

* **Diagnosis and Troubleshooting of Devices and Systems:** This class comprises systems that deduce faults and suggest corrective actions for a malfunctioning device or process. Medical diagnosis was one of the first knowledge areas to which ES technology was applied, but diagnosis of engineered systems quickly surpassed medical diagnosis.
* **Planning and Scheduling :** Systems that fall into this class analyze a set of one or more potentially complex and interacting goals in order to determine a set of actions to achieve those goals. This class has great commercial potential, which has been recognized. Examples involve airline scheduling of flights, personnel, and gates; manufacturing job-shop scheduling; and manufacturing process planning.
* **Configuration of Manufactured Objects from Subassemblies: Configuration,** Configuration applications were pioneered by computer companies as a means of facilitating the manufacture of semi-custom minicomputers. The technique has found its way into use in many different industries, for example, modular home building, manufacturing, and other problems involving complex engineering design and manufacturing.
* **Financial Decision Making:** The financial services industry has been a vigorous user of expert system techniques. For example, Insurance companies have used expert systems to assess the risk presented by the customer and to determine a price for the insurance.
* **Knowledge Publishing:** This is a relatively new, but also potentially explosive area. The primary function of the expert system is to deliver knowledge that is relevant to the user's problem, in the context of the user's problem.
* **Process Monitoring and Control:** Systems falling in this class analyze real-time data from physical devices with the goal of noticing anomalies, predicting trends, and

controlling for both optimality and failure correction. Examples of real-time systems that actively monitor processes can be found in the steel making and oil refining industries.

* **Design and Manufacturing:** These systems assist in the design of physical devices and processes, ranging from high-level conceptual design of abstract entities all the way to factory floor configuration of manufacturing processes

PROCESS IN EXPERT SYSTEM WORKING

Expert system works with the support of two types of interfaces, these are user interface and procedure interface. The user interface and the procedure interface are two important functions in the information collection process.

The user interface

The function of the user interface is to present questions and information to the operator and supply the operator's responses to the inference engine. Any values entered by the user are received and interpreted by the user interface. Some responses are restricted to a set of possible legal answers, others are not. The user interface checks all responses to insure that they are of the correct data type. Any responses that are restricted to a legal set of answers are compared against these legal answers. Whenever the user enters an illegal answer, the user interface informs the user that answer was invalid and prompts to correct it.

Procedure node interface

The function of the Procedure node interface is to receive information from the Procedures coordinator and create the appropriate procedure call. The ability to call a procedure and receive information from that procedure can be viewed as simply a generalization of input from the external world. While in some earlier expert systems external information has been obtained, that information was obtained only in a predetermined manner so only certain information could actually be acquired. This expert system, is permitted to invoke any Procedure allowed on its host system. This makes the expert system useful in a much wider class of knowledge domains than if it had no external access or only limited external access.

The knowledge that is represented in the system appears in the Rulebase. In the Rulebase described in the cross-referenced applications, there are basically four different types of objects, with associated information present.

1. ***Classes*** **:** These are questions asked to the user.
2. ***Parameters*** **:** A Parameter is a place holder for a character string which may be a variable that can be inserted into a Class question at the point in the question where the Parameter is positioned.
3. ***Procedures*** **:** These are definitions of calls to external Procedures.
4. ***Rule Nodes*** **:** The inferencing in the system is done by a tree structure which indicates the Rules or logic which mimics human reasoning. The nodes of these trees are called RULE nodes.

It has been found that the type of information that is solicited by the system from the user by means of questions or classes should be tailored to the level of knowledge of the user. In many applications, the group of prospective uses is nicely defined and the knowledge level can be estimated so that the questions can be presented at a level which corresponds generally to the average user. However, in other applications, knowledge of the specific domain of the expert system might vary considerably among the group of prospective users.

Working of Expert System

Expert Systems consist of

* knowledge base (facts)
* production rules ("if.., then..")

* inference engine (controls how "if.., then.." rules are applied towards facts)

Actually there are two methods to make conclusions.

Method Name	*Short Explanation*	*Use*	*Example Systems*
Forward chaining	facts driven	can find new ideas	CLIPS, Jess
Backward chaining	hypothesis driven	usually used for diagnosis	Prolog, Mycin

APPLICATION OF EXPERT SYSTEM IN LIBRARY AND INFORMATION SYSTEM

Having achieved remarkable success in both industry and commerce, expert systems are now denuding the frontiers of library and information systems. There is a significant surge to activities on assessing and debating the likely impact of expert systems on library and information profession.

Brooks and Jones observed that subdivision of library and information science field are suitable for application of expert systems as they possess clear cut boundary with lots of public and heuristics knowledge which can be identified, elicited, codified and finalized in the knowledge base. Their claim is vindicated by a study which has identified 139 expert system projects in the LIS domain. Expert systems are usually grouped by the sub domain for example, online information service, reference work, cataloguing, indexing etc. A break of availability of expert systems is given below:

Reference work	32 percent
Online retrieval	28 percent
Indexing, cataloguing, classification	12 percent
Library management application	7 percent
Abstracting	1 percent

EXPERT SYSTEMS RESEARCH

The primary goal of expert systems research is to make expertise available to decision makers and technicians who need answers quickly. There is never enough expertise to go around - certainly it is not always available at the right place and the right time. Portable with computers loaded with in-depth knowledge of specific subjects can bring decade's worth of knowledge to a problem. The same systems can assist supervisors and managers with situation assessment and long-range planning. Many small systems now exist that bring a narrow slice of in-depth knowledge to a specific problem, and these provide evidence that the broader goal is achievable. These knowledge-based applications of artificial intelligence have enhanced productivity in business, science, engineering, and the military. With advances in the last decade, today's expert systems clients can choose from dozens of commercial software packages with easy-to-use interfaces. Each new deployment of an expert system yields valuable data for what works in what context, thus fueling the AI research that provides even better applications.

ADVANCED RESEARCH IN KNOWLEDGE-BASED SYSTEMS: INVENTING THE NEXT GENERATION

The basic categories of research in knowledge-based systems include: knowledge representation, knowledge use (or problem-solving), and knowledge acquisition (i.e., machine learning and discovery).

Knowledge Representation

In knowledge representation, the key topics are concepts, languages, and standards for knowledge representation. There are many issues involved in scaling up expert systems: defining the problems encountered in the pursuit of large knowledge bases; developing the infrastructure for building and sharing large knowledge bases; and actually accumulating a large body of knowledge, for example, common sense knowledge or engineering and technical knowledge.

Knowledge Use

Knowledge use, or problem-solving, research efforts

involve the development of new methods for different kinds of reasoning, such as analogical reasoning, reasoning based on probability theory and decision theory, and reasoning from case examples. The first generation of expert systems was characterized by knowledge bases that were narrow and, hence, performance that was brittle. When the boundary of a system's knowledge was traversed, the system's behavior went from extremely competent to incompetent very quickly. To overcome such brittleness, researchers are now focusing on reasoning from models, principles and causes. Thus, the knowledge-based system will not have to know everything about an area, as it were, but can reason with a broader base of knowledge by using the models, the principles, and the causation.

Knowledge Acquisition

The quest for a large knowledge base boils down to the problem of access to distributed knowledge bases involving multiple expert systems and developers. The effort to develop the infrastructure needed to obtain access is a research area called knowledge sharing. The goal of the knowledge sharing research is to overcome the isolation of first-generation expert systems, which rarely interchanged any knowledge.

Other Areas of Investigation

A major issue of expert systems research involves methods for reasoning with uncertain data and uncertain knowledge. One of the most widely adopted methods is called "fuzzy logic" or "fuzzy reasoning," especially in Japan, where fuzzy reasoning is the object of much research attention and much scrutiny on the part of researchers. Very lately, there has come on the scene the research topic of neural networks — networks of distributed components operating in parallel to make classification decisions. The links between neural networks technology and expert system technology are being explored. Finally, research is underway to explore the use of new parallel computing methods in the implementation of expert systems and advanced knowledge-based systems. The new wave of computing is multiprocessor technology. The question is, what will be the impact of such high-performance parallel computing activities on expert system techniques.

REFERENCES

1. Anonymous. *Dictionary of Computing*, 1986, New York: Oxford University Press.
2. Bishop, Peter. *Fifth Generation Computers Concepts, Implementations & Uses*, 1986, Chichester, England: Ellis Horwood Ltd.
3. Brule, James F. *Artificial Intelligence: Theory, Logic and Application*, 1986, Blue Ridge Summit, PA: TAB Books.
4. http://www.expertsystem.net/?lang=1
5. http://www.bbc.co.uk/scotland/education/bitesize/standard/computing/gp_packages/index.shtml
6. http://www.britannica.com/eb/article-219098
7. Edwards, Alex and Connell, N.A.D. *Expert Systems in Accounting*, 1989, Herfordshire, UK: Prentice Hall International (UK) Ltd.
8. Forsyth, Richard, *Expert Systems: Principles and Case Studies*, 1984, London: Chapman and Hall Computing.
9. http://www.exsys.com/demomain.html
10. http://www.forbes.com/global/1998/1130/0118096a.html
11. http://www.openclinical.org/dss.html#firstGen
12. http://www.archive.org/details/CC1024_artificial_intelligence
13. Harmon, Paul and King, David. *Expert Systems: Artificial Intelligence in Business*. 1985, New York: Wiley.
14. Liebowitz, Jay, *Introduction to Expert Systems*, 1988, Santa Cruz, CA: Mitchell Publishing, Inc.
15. http://www.aaai.org/Library/Magazine/vol17.php#Winter
16. http://www.aaai.org/AITopics/classic/buchanan.html
17. http://www.aaai.org/Library/Magazine/vol10.php#Spring
18. Michaelsen, Robert H.; Michie, Donald and Boulanger, Albert. "The Technology of Expert Systems" *Byte*; April 1985; 10(4): pp. 303-312.
19. Rich, Elaine and Knight, Kevin. *Artificial Intelligence Second Edition*. 1991, New York: McGraw-Hill.
20. Waterman, Donald A. *A Guide to Expert Systems*, 1986, Reading, MA: Addison-Wesley.
21. Winston, Patrick H. and Prendergast, Karen A. (Editors). *The AI Business: Commercial Use of Artificial Intelligence*, 1984, Cambridge, MA: The MIT Press.
22. http://www.expertise2go.com/webesie/tutorials/ESIntro/

23. http://www.insurancenetworking.com/protected/article.cfm?articleId=4331
24. http://www.aaai.org/Library/Magazine/vol03.php#Summer
25. http://www.epa.gov/oppt/cahp/pubs/can.htm
26. http://www.fcw.com/article97625-02-12-07
27. http://citeseer.ist.psu.edu/feigenbaum92personal.html
28. Brooks, H M. Expert Sytems and Intellignet Inforamtion Retrieval. Information Processing Management, 1987, 23(4), pp. 367-92.
29. Jones, K and Span, K. Information Retrieval. In, Encyclopaedia of Artificial Intelligence. Ed. By Shapiro, S C John. New York: Wiley & Sons. New York, 1987, vol"; 419-21.
30. Poulter, A and Others. LIS Professioanls as knowledge engineers. In: Annual Review of Information Science and Technology
31. Dubey, Y P. Expert Systems and Their application in Library and Information Science. DESIDOC Bulletin of Infomration Technology, 1996, 6(4), pp. 37.
32. *http://www.wtec.org/loyola/kb/c3_s3.htm*

Appendix 1
Information Technology Glossary

16-bit

Software for MS-DOS or Microsoft Windows originally ran on the 16-bit Intel 8088 and 80286 microprocessors.

32-bit

32-bit programs are written for the Intel 80386 and more recent processors, and allow greater speed of execution.

Access Point

A name, term, code, etc., under which a book or other item may be searched and identified in the catalog, for example: author, title, subject, call number.

API - Application Programming Interface

An interface between the operating system and application programs, which includes the way applications communicate with the operating system, and the services the operating system makes available to programs.

Architecture

Design, or the way components fit together. It can be used to describe any system, as in "software architecture" or "network architecture".

Authentication

Verification of identity for security purposes. Passwords and digital signatures are forms of authentication.

Authority Control

A method to control the number of similar headings (usually subjects or authors) in the library catalog. Multiple entries are cross-referenced to a single entry using authority control.

Automation

Automatic, as opposed to human, operation or control of a process, equipment or a system; or the techniques and equipment used to achieve this. In libraries, automation refers to the process of automating functions such as circulation, cataloging, or acquisitions.

Barcode

A printed horizontal strip of vertical bars used for identifying specific items or users. The codes, which represent numerical data, are read by a bar code reader and interpreted via software or hardware decoders. In libraries, barcodes are affixed to both books and library cards to assist in circulation and collection control.

Boolean Operators

Words such as AND, OR, and NOT used to broaden or narrow a search. For example, **cats AND dogs** narrows the search to titles about both cats and dogs, while **cats OR dogs** broadens the search.

Browse Searching

A feature of a catalog which allows the user to search by the first word in a field, not a word appearing anywhere in the field. Compare to *Keyword Searching.*

Browser

See *Web Browser*

CGI Script

A computer program, most frequently written in C, Perl, or a shell script, that uses the Common Gateway Interface (CGI) standard to provide an interactive interface between a user or an application and a Web server. CGI script is most commonly used to develop forms that allow users to submit information to a Website.

Character Set

A standard collection of characters. A character set may include letters, digits, punctuation, control codes, graphics, mathematical symbols, and other signs. Each character in the set is represented by a unique character code, which is a binary number used for storage and transmission.

Chat Room

An online space where two or more people can input (usually via keyboard) messages to a group of people who are logged in at the same time. Compare to *Message Forum*.

Client

A computer system or process that requests a service of another computer system or process (see *Server*). For example, a workstation requesting a file from a server is a client of the file server.

Client/Server

A common type of distributed system in which software is split between server tasks and client tasks. A client sends requests to a server asking for information or action, and the server responds. There may be either one centralized server or many distributed ones.

Codabar

A specific type of bar code used in libraries, blood banks, the overnight package delivery industry, and a variety of other information processing applications. It is also known as USD-4, NW-7, and 2 of 7 code. See *Barcode*.

Command Line Interface

A means of communication between a program and its user in which commands are input with a keyboard and are interpreted and executed by the program. Results are output as text or graphics on the computer screen. Compare to *Graphical User Interface*.

Cookie

Data created on your own computer when you access certain Web pages. Usually the cookie keeps information about your preferences and reuses this information when you access that site again, thus "recognizing" repeat visitors. Cookies cannot be used to access any other data on your computer, nor can they determine your e-mail address or identity. You can set your browser to reject all cookies or to warn you before accepting cookies, but sites that provide personalized services, such as stock trading sites or shopping carts, won't work if you don't accept cookies.

Database

A structured set of data, generally accessed via a software program. A simple database might be a single file containing many records, each of which contains the same set of fields, such as a series of companies with name, address, phone, and contact fields for each one.

Database Management System (DBMS)

Complex set of programs that control the organization, storage and retrieval of data for many users. Data is organized in fields, records and files. Examples include Oracle, Sybase, and Datacom.

Dedicated Line

A permanent connection between computers using telephone, ISDN, or other types of lines.

Dialup

A temporary, as opposed to dedicated, connection

between machines established over a telephone line using modems. A patron might use a dialup connection from home (using a personal computer, modem, and telephone line) to dial into and use the library catalog.

Digital Document

Books, articles, papers, etc., that can be accessed via a computer.

Digital Media

Media in various digital forms, including scanned images, audio, video, multimedia, drawings and intelligent documents.

Domain Name System (DNS)

The unique name of a collection of computers connected to a network such as the Internet. A replicated, distributed data query service for looking up host IP addresses based on host names. The DNS is hierarchical, consisting of domains, subdomains, sites, and hosts. Unique names are formed from smallest to largest, and are of the form *user@host.site.subdomain.domain*, where *host* and *site* are often optional. On the Internet, domain names typically end with a suffix denoting the type of site:

* **.com** (commercial)
* **.edu** (educational)
* **.net** (network operations)
* **.gov**(US government)
* **.mil** (US military)
* **.org** (organization)
* **.us** (United States)
* **.ca** (Canada)
* **.uk** (United Kingdom)
* **.au** (Australia
* **.cz** (Czech Republic)
* **.xx**(where *xx* refers to another country's two-letter abbreviation)

Download

To copy a file from an external computer to a local one. You may download software updates from a vendor to your library system via the Internet.

Dublin Core

A minimal set of metadata elements used to describe networked information resources which aids users in locating specific items.

Dumb Terminal

An output device (not a computer) that contains no internal microprocessor; a display monitor with no processing capabilities.

EAD

Encoded Archival Description is the emerging standard for archival finding aids and is used by Library of Congress and other research institutions. Finding aids are inventories, registers, indexes, or guides that provide detailed information about specific collections. Standardizing the format of these tools makes it easier to display them on a network.

EDIFACT (ISO 9735)

Electronic data interchange (EDI) refers to any electronic commerce standard that defines a protocol for exchange of business data between software applications. EDIFACT (EDI for Administration, Commerce, and Transport) is the international standard for EDI.

Encryption

The process of scrambling a message so that a key, held only by authorized recipients, is needed to unscramble and read the message.

Ethernet

Computers on the Internet that use the TCP/IP protocols are frequently connected to the Net over an Ethernet link. Ethernet supports communications at 10 mbps over several types of wiring. An Ethernet card connects an individual computer to a network.

Extranet

An extended int*ra*net connecting not only internal personnel, but also select customers, suppliers, and strategic partners. Compare to Internet and Intranet.

FAQ — Frequently Asked Questions

Web page or text file with common questions about various topics along with the answers. Especially useful for first-time visitors to a site or new participants in a newsgroup or listserv.

Filtering software

A program to block access to certain Web pages. Depending on the program, it may block certain specified sites, all sites with certain words, or it may use more sophisticated criteria to determine which pages to block. See the LibraryHQ's Internet Filtering and Blocking page.

Firewall

A gateway used to protect a server or a network from unauthorized access. A firewall generally consists of both hardware and software components.

FTP

See *Protocols*

Graphical User Interface (GUI)

The use of pictures rather than just words to represent the input and output of a program. The program displays icons, buttons, dialog boxes etc. in its windows on the screen; the user controls it by moving a mouse or pointer on the screen, selecting objects by pressing buttons on the mouse. Compare to *Command Line Interface*.

HTML

Hypertext Markup Language is the set of codes (or tags) inserted in a file intended for display on a Web browser. The codes tell the Web browser how to display the words and images on the page.

HTTP

See *Protocols*

Hypertext

Text on a Web page that, when clicked on, jumps the reader to another page or image.

Integrated Library System (ILS)

An automated library system in which all of the functional modules (acquisitions, circulation, cataloging, serials, an OPAC, etc.) are accessible online and share a common machine-readable database. An automated library system is considered *integrated* only if all library functions are processed against a single, master bibliographic file.

Imported/Exported Bibliography Records

Because of the standards established for the creation and transferal of library bibliographic records, libraries can exchange records through import and export.

Interface

In hardware, an interface is a connector used to link devices. In software, it allows communication between two software systems or between people and systems. In the automation field, interface refers to the method by which users can access the automated library system. See *Graphical User Interface*.

Internet

A network of networks; a group of networks interconnected via routers. The Internet is the world's largest network.

Internet Service Provider

A company that provides other companies or individuals with access to, or a presence on, the Internet.

Intranet

A network of computers and servers accessed via Web browsers, but maintained within an organization or company, and not generally accessible to those outside the organization.

ISBN (ISO 2108)

International Standard Book Number: a unique numerical 10-digit identifier for a published title. It helps to ensure more efficient ordering, inventory control, and accounting.

ISSN (ISO 3297)

International Standard Serial Number: identifies serial publications including monographic series. In 1979 the ISSN replaced the ISBN for serials.

Java

A programming language designed so that its object modules can run on many different platforms. Java is often used to create *applets*, or Web-based programs that run when a user accesses the page or clicks on a certain area.

JavaScript

A scripting language used to create *scripts*, or small programs embedded in Web pages that run when a user accesses the page or clicks on a certain area.

Keyword Searching

A keyword search looks for specific words or terms that occur anywhere in a field (title, subject, heading, contents, etc.). Keyword searching is useful when you have incomplete information. Compare to *Browse Searching.*

LAN – Local Area Network

A geographically limited data communications network that connects several computers. Library automation systems require a LAN to enable all users to access the same database.

Legacy System

An older computer system or program that is still valuable

to the user, but may not be compatible with newer systems.

Licensing

An automation system requires the library to obtain a user license, as with any type of software. Licenses are usually purchased as either a site license, which allows the library to install the automation software on as many computers desired within a single building, or on a per-computer or per-network basis.

Listserv

A Listserv is a program that automatically redistributes e-mail to names on a mailing list. Users can subscribe, unsubscribe, and send messages to everyone on the list by sending notes to a specific email address.

MARC - (Machine-Readable Cataloging)

MARC refers to (1) a computer record structure, (2) a set of tags and indicators to identify parts of the record, (3) the level of cataloging information contained in the Library of Congress's MARC records, and (4) the body of records distributed by the Library of Congress MARC Distribution Service.

Message Forum

Also called a Message Board. An online space where people can leave messages that become visible to anyone who accesses the forum. Compare to *Chat Room.*

Module/Application Module

A module is a software segment that performs a specific function, such as acquisitions or circulation. Automation system vendors may sell modules separately, though circulation and cataloging modules are often sold together, with add-on modules as possible extra purchases.

NetPC, NetworkPC

A computer with minimal memory, disk storage and processor power designed to connect to and work via a network.

Network

A group of interconnected computers, including the hardware, software, and cabling used to connect them.

Newsgroups

see Usenet Newsgroups

NISO

A nonprofit association accredited as a standards developer by the American National Standards Institute, the national clearinghouse for voluntary standards development in the U.S. NISO has developed standards for Information Retrieval (Z39.50), 12083 (an SGML Tool), Z39.2 (Information Interchange Format), Codes for Languages and Countries, and Z39.18 (Scientific and Technical Reports).

OPAC

Online Public Access Catalog: provides access to the library's holdings via a computer monitor, replacing the traditional card catalog. May also be called a PAC (Public Access Catalog).

Operating Systems (OS)

The low-level software on a computer that schedules tasks, allocates storage, handles the interface to peripheral hardware, and presents a default interface to the user when no application program is running. Examples: UNIX, Windows98, Windows NT.

Protocols

Sets of rules or standards that let computers communicate over the Internet. **HTTP** (Hypertext Transfer Protocol) allows transfer of Web pages via a browser. **FTP** (File Transfer Protocol) allows transfer of files through the Internet from one computer to another.

Push Technology

Internet technology that allows information to be delivered or "pushed" directly to a user who subscribes to it.

PointCast, Yahoo, and other services provide news to users who can customize their news reports by selecting topics like sports, world news, and business that are entered into a database.

Query

A user's request for information from a database or search engine. A query is a search string entered by a patron when searching for a particular library item or subject in the library's catalog, or OPAC.

RDBMS

Relational Database Management System. See *Relational Database.*

Record Conversion

Record conversion is an aspect of retrospective conversion, in which book records on a tape or disk are converted to MARC format for use in an automation system. Conversion is usually processed by the automation system or by an outside vendor specializing in retrospective conversion.

Relational Database

A database in the form of tables having rows and columns to show the relationships between items. If data is changed in one table, it will be changed in all related tables. See *Database.*

Remote Access

Service allowing users away from the server or network to access these resources from remote locations.

Retrospective Conversion

Retrospective conversion is the process by which libraries convert a shelflist (such as a card catalog) to a searchable, computerized database. This database can then be used as the backbone of an automation system.

Request For Information (RFI)

A formal request that a library prepares and sends to vendors for information on their automation systems.

Request For Proposal (RFP)

A formal open bid document that a library prepares to obtain a customized response from a vendor. Responses may include pricing, enhancements, implementation plan, and timeline. An RFP is usually the precursor to a legal contract between vendor and library.

Server

A computer that provides some service for other computers connected to it via a network. The library's database containing all book records is located on a server so that several client machines (OPACs) can access the files. See *Client/Server.*

Site Preparation

The process of installing wiring, cabling, and other support hardware, in addition to planning the placement of computers, terminals, and networks within the library.

System Upgrade

A new or better version of hardware or software, or the act of developing or installing a new version. An automation system upgrade might be the change between a command-line based system and a GUI-based or Web-based automation system.

TCP/IP

Transmission Control Protocol/Internet Protocol, the suite of communications protocols used to connect computers on the Internet.

Thin Client

A thin client is a machine that only functions when connected to a central server, as compared to a networked PC, which can store and process information on its hard drive. The function of a thin client is to centralize data transmission and storage in a LAN environment.

Truncation

In searching, truncation means that the system will allow

wildcards to extend a search term. For example, a truncated term with a wildcard, such as *app**, will retrieve all entries that begin with *app*, from *apples* to *application*. See *Wildcard*.

Turnkey

A complete hardware and software computer system customized for a particular application, such as library automation. Turnkey systems are fully operational, requiring no additional components or modifications.

Unicode

A 16-bit, language-independent character set that enables representation of all of the characters commonly used in information processing.

UniMARC

An international MARC format that accepts records created in any of the more than 20 MARC formats, thus facilitating conversion of records among them.

UNIX

Widely used multi-user general-purpose computer operating system.

User interface

The aspects of a computer system or program that can be perceived by the human user, and the commands and mechanisms used to control its operation and input data. *See Graphical User Interface* and *Command Line Interface.*

Usenet Newsgroups

Usenet is a collection of messages on various subjects that are posted to servers on a worldwide network. Each subject collection is called a newsgroup. Most newsgroups are hosted on Internet-connected servers, but they can also be hosted from servers that are not part of the Internet.

URL

Uniform Resource Locator: the address a computer uses to locate documents on the Internet.

Video Display Unit

A device incorporating a display monitor, a keyboard and a serial port. The most common VDUs, "dumb terminals, " are used primarily to send and receive information from a central computer.

Virtual Library

A virtual (as opposed to physical) collection existing only as digital files accessible via computer, not as books or tangible media in a library building.

Web Browser

Software program that allows you to access Web pages on the *Internet*, an *Intranet*, or an *Extranet.* The two most popular browsers are Microsoft's Internet Explorer and Netscape Navigator.

Wide Area Network (WAN)

A network, usually constructed with serial lines, extending over distances greater than one mile. A library system may use a WAN to connect branch libraries to the greater automation system.

Wildcard

A special character such as *, $, or ? that can replace any character or characters in a string. Wildcards are used in catalog searches to extend a search term. See *Truncation*.

WindowsNT

Microsoft's 32-bit operating system designed for high-end workstations, servers and corporate networks.

Wizard

An interactive help utility that guides the user through a potentially complex task.

X12

An Electronic Data Interchange standard — the American equivalent of UN/EDIFACT and various other national and industry EDI messaging protocols.

Z39.20

A NISO (National Information Standards Organization) standard for measuring the extent of price changes on a periodic basis for a variety of library materials including hardcover trade and technical books, paperback books, periodicals, and microforms.

Z39.50

A NISO (National Information Standards Organization) standard for information retrieval that allows any library using a Z39.50-compliant automated library system to access remote library collections. Z39.50 specifies a query/response protocol between a client and a server. See LibraryHQ's Z39.50 page.

Appendix: 2
Data Mining Terms Glossary

analytical model

A structure and process for analyzing a dataset. For example, a decision tree is a model for the classification of a dataset.

anomalous data

Data that result from errors (for example, data entry keying errors) or that represent unusual events. Anomalous data should be examined carefully because it may carry important information.

artificial neural networks

Non-linear predictive models that learn through training and resemble biological neural networks in structure.

CART

Classification and Regression Trees. A decision tree technique used for classification of a dataset. Provides a set of rules that you can apply to a new (unclassified) dataset to predict which records will have a given outcome. Segments a dataset by creating 2-way splits. Requires less data preparation than CHAID.

CHAID

Chi Square Automatic Interaction Detection. A decision tree technique used for classification of a dataset. Provides a set of rules that you can apply to a new (unclassified) dataset to predict which records will have a given outcome. Segments a dataset by using chi square tests to create multi-way splits. Preceded, and requires more data preparation than, CART.

classification

The process of dividing a dataset into mutually exclusive groups such that the members of each group are as "close" as possible to one another, and different groups are as "far" as possible from one another, where distance is measured with respect to specific variable(s) you are trying to predict. For example, a typical classification problem is to divide a database of companies into groups that are as homogeneous as possible with respect to a creditworthiness variable with values "Good" and "Bad."

clustering

The process of dividing a dataset into mutually exclusive groups such that the members of each group are as "close" as possible to one another, and different groups are as "far" as possible from one another, where distance is measured with respect to all available variables.

data cleansing

The process of ensuring that all values in a dataset are consistent and correctly recorded.

data mining

The extraction of hidden predictive information from large databases.

data navigation

The process of viewing different dimensions, slices, and levels of detail of a multidimensional database. See OLAP.

data visualization

The visual interpretation of complex relationships in multidimensional data.

data warehouse

A system for storing and delivering massive quantities of data.

decision tree

A tree-shaped structure that represents a set of decisions. These decisions generate rules for the classification of a dataset. See CART and CHAID.

dimension

In a flat or relational database, each field in a record represents a dimension. In a multidimensional database, a dimension is a set of similar entities; for example, a multidimensional sales database might include the dimensions Product, Time, and City.

exploratory data analysis

The use of graphical and descriptive statistical techniques to learn about the structure of a dataset.

genetic algorithms

Optimization techniques that use processes such as genetic combination, mutation, and natural selection in a design based on the concepts of natural evolution.

linear model

An analytical model that assumes linear relationships in the coefficients of the variables being studied.

linear regression

A statistical technique used to find the best-fitting linear relationship between a target (dependent) variable and its predictors (independent variables).

logistic regression

A linear regression that predicts the proportions of a

categorical target variable, such as type of customer, in a population.

multidimensional database

A database designed for on-line analytical processing. Structured as a multidimensional hypercube with one axis per dimension.

multiprocessor computer

A computer that includes multiple processors connected by a network. See parallel processing.

nearest neighbor

A technique that classifies each record in a dataset based on a combination of the classes of the k record(s) most similar to it in a historical dataset (where k ^{3} 1). Sometimes called a k-nearest neighbor technique.

non-linear model

An analytical model that does not assume linear relationships in the coefficients of the variables being studied.

OLAP

On-line analytical processing. Refers to array-oriented database applications that allow users to view, navigate through, manipulate, and analyze multidimensional databases.

outlier

A data item whose value falls outside the bounds enclosing most of the other corresponding values in the sample. May indicate anomalous data. Should be examined carefully; may carry important information.

parallel processing

The coordinated use of multiple processors to perform computational tasks. Parallel processing can occur on a multiprocessor computer or on a network of workstations or PCs.

predictive model

A structure and process for predicting the values of specified variables in a dataset.

prospective data analysis

Data analysis that predicts future trends, behaviors, or events based on historical data.

RAID

Redundant Array of Inexpensive Disks. A technology for the efficient parallel storage of data for high-performance computer systems.

retrospective data analysis

Data analysis that provides insights into trends, behaviors, or events that have already occurred.

rule induction

The extraction of useful if-then rules from data based on statistical significance.

SMP

Symmetric multiprocessor. A type of multiprocessor computer in which memory is shared among the processors.

terabyte

One trillion bytes.

time series analysis

The analysis of a sequence of measurements made at specified time intervals. Time is usually the dominating dimension of the data.